Computational Molecular Science and Engineering Forum

Held at the 2024 AIChE Annual Meeting

San Diego, California, USA
27-31 October 2024

ISBN: 979-8-3313-1656-3

Printed from e-media with permission by:

Curran Associates, Inc.
57 Morehouse Lane
Red Hook, NY 12571

Some format issues inherent in the e-media version may also appear in this print version.

Printed with permission by Curran Associates, Inc. (2025)

For permission requests, please contact AIChE
at the address below.

AIChE
120 Wall Street, FL 23
New York, NY 10005-4020

Phone: (800) 242-4363
Fax: (203) 775-5177

www.aiche.org

Additional copies of this publication are available from:

Curran Associates, Inc.
57 Morehouse Lane
Red Hook, NY 12571 USA
Phone: 845-758-0400
Fax: 845-758-2633
Email: curran@proceedings.com
Web: www.proceedings.com

TABLE OF CONTENTS

APPLICATIONS OF MOLECULAR MODELING TO STUDY INTERFACIAL PHENOMENA

AUTOMATED MOLECULAR AND MATERIALS DISCOVERY: INTEGRATING MACHINE LEARNING, SIMULATION, AND EXPERIMENT

MACHINE LEARNING FOR SOFT AND HARD MATERIALS

PLENARY SESSION: COMPUTATIONAL MOLECULAR SCIENCE AND ENGINEERING FORUM (INVITED TALKS)

POSTER SESSION: COMPUTATIONAL MOLECULAR SCIENCE AND ENGINEERING FORUM

RECENT ADVANCES IN MOLECULAR SIMULATION METHODS

RECENT ADVANCES IN MULTISCALE METHODOLOGIES

SOFTWARE ENGINEERING IN AND FOR THE MOLECULAR SCIENCES

SPOTLIGHTS IN THERMODYNAMICS AND COMPUTATIONAL MOLECULAR SCIENCE (INVITED TALKS)

INDUSTRIAL APPLICATIONS OF COMPUTATIONAL CHEMISTRY AND MOLECULAR SIMULATION

SPECIAL SESSION IN HONOR OF PROF. SHARON GLOTZER'S 60TH BIRTHDAY I (INVITED TALKS)

APPLICATIONS OF MOLECULAR MODELING TO STUDY INTERFACIAL PHENOMENA II

MACHINE LEARNING FOR SOFT AND HARD MATERIALS II

AUTOMATED MOLECULAR AND MATERIALS DISCOVERY: INTEGRATING MACHINE LEARNING, SIMULATION, AND EXPERIMENT II

RECENT ADVANCES IN MOLECULAR SIMULATION METHODS II

SPECIAL SESSION IN HONOR OF PROF. SHARON GLOTZER'S 60TH BIRTHDAY II (INVITED TALKS)

393a Molecular Simulation Study of Gas Adsorption and Transport in Room Temperature Ionic Liquid Membranes Supported By Carbon Nanotubes

Gaurab Sarkar, Andrew J. Schultz, Jeffrey R. Errington

Development of robust strategies to capture carbon dioxide from flue gas emissions in power plants represents a key step in addressing factors that contribute to climate change. Membranes featuring room temperature ionic liquids (RTILs) have been identified as potentially effective media for capturing carbon dioxide. The very low vapor pressures and tunable nature of RTILs make them attractive media for membranes. The Yu group at the University at Buffalo has recently pursued an approach in which the RTIL is supported by a mesh of single-walled carbon nanotubes (SWCNT). The approach results in a membrane with nano-sized features. The mesh structure potentially provides enhanced mechanical, transport, and selectivity characteristics. In this work, we use molecular simulation tools to study gas adsorption and transport properties within mesh-based RTIL structures, and compare the performance of these systems to bulk RTIL systems. We first discuss generation of a molecular model that aims to capture the essential features of the mesh structure. We next describe the Monte Carlo simulation protocol developed to study gas adsorption. Specifically, we employ a strategy in which the temperature, pressure, numbers of molecules of RTIL and SWCNT, and the activities of relevant gases are fixed. The approach enables us to capture the adsorption characteristics of various model flue gases over a range of pressures and compositions. Finally, we present results for the adsorption and diffusion characteristics of select systems over a broad range of conditions. These characteristics are then linked to the manner in which the RTIL and gases organize within the nano-confined space.

393b Formation of Nanoscale Ionic Liquid Clusters to Dictate the Electrostatic Screening Length Via Molecular Dynamics

Lisa Je, Victor Zavala, Reid Van Lehn

Ionic liquids (ILs), such as 1-ethyl-3-methylimidazolium tetrafluoroborate (EMim BF_4), have been scrutinized as viable electrolytes for electrochemical reactions due to their highly tunable cation/anion properties, intrinsic conductivity, wide electrochemical stability window, low volatility, and strong electrostatic interactions [1-3]. When mixed with organic solvents, ILs have been proposed to form nanoscale clusters that lead to unexpected effects on electrostatic screening. For example, previous experimental studies have identified an apparent nonmonotonic variation in the Debye screening length with IL concentration in contrast to conventional expectations from Debye-Hückel theory [4]. Similarly, recent experimental studies of electrocatalytic CO_2 reduction have identified a similar nonmonotonic dependence of electrocatalytic activity on IL concentration in IL-solvent mixtures, which may be related to nanoscale cluster formation. However, the molecular structure of ILs mixed with organic solvents remains poorly understood. Inhibiting mechanistic interpretation of these behaviors and prevents the rational design of ILs [5-7]. Molecular dynamics (MD) simulations can address this problem by modeling the dynamic formation of IL nanostructures that is difficult to observe via experiments.

In this work, we performed all-atom MD simulations to quantify concentration-dependent changes in EMIM-BF_4 structure by modeling very dilute concentrations (0.001 mol fraction) to concentrated IL solutions (0.25, 0.5, 1 mol fraction) in acetonitrile. These MD simulations allow us to index individual ions and trace their movements as they form nanoclusters of different charge. Simulation analysis revealed a nonmonotonic trend in the both the formation of neutral clusters (*i.e.* clusters with no net charge) and mobile charge carriers (*i.e.* individual ions and clusters with a non-zero charge) with IL concentration, in broad agreement with nonmonotonic trends in electrostatic properties observed experimentally [7]. Notably, the average number of neutral clusters and mobile charged clusters reach a maximum that coincides with a maximum current density in previously performed electrocatalytic studies [7]. We further varied the solvent chemical identity ranging from a low dielectric constant (acetone), intermediate (dimethyl sulfoxide), and high (propylene carbonate) to predict changes in cluster behavior. Next, we introduce a silver electrode to the IL to observe the impact of the interface on the IL nanostructure. Together, these simulations provide molecular details regarding how these three tuning knobs (concentration, solvent, electrode) affect the formation of nanoscale IL clusters that dictate the electric double layer and potentially the electrocatalytic activity based on comparisons to experiments [7].

[1] K. Shimizu, C. E. S. Bernardes, and J. N. Canongia Lopes, 'Structure and Aggregation in the 1-Alkyl-3-Methylimidazolium Bis(trifluoromethylsulfonyl)imide Ionic Liquid Homologous Series,' *The Journal of Physical Chemistry B,* vol. 118, no. 2, pp. 567-576, 2014.

[2] K. Motobayashi, Y. Maeno, and K. Ikeda, 'In Situ Spectroscopic Characterization of an Intermediate of CO2 Electroreduction on a Au Electrode in Room-Temperature Ionic Liquids,' *The Journal of Physical Chemistry C,* vol. 126, no. 29, pp. 11981-11986, 2022.

[3] B. A. Rosen *et al.*, 'In Situ Spectroscopic Examination of a Low Overpotential Pathway for Carbon Dioxide Conversion to Carbon Monoxide,' *The Journal of Physical Chemistry C,* vol. 116, no. 29, pp. 15307-15312, 2012.

[4] A. M. Smith, A. A. Lee, and S. Perkin, 'The Electrostatic Screening Length in Concentrated Electrolytes Increases with Concentration,' *The Journal of Physical Chemistry Letters,* vol. 7, no. 12, pp. 2157-2163, 2016.

[5] B. A. Rosen *et al.*, 'Ionic Liquid-Mediated Selective Conversion of CO2 to CO at Low Overpotentials,' (in English), *Science,* Article vol. 334, no. 6056, pp. 643-644, Nov 2011, doi: 10.1126/science.1209786.

[6] M. A. Gebbie *et al.*, 'Long range electrostatic forces in ionic liquids,' *Chemical Communications,* vol. 53, no. 7, pp. 1214-1224, 2017.

[7] B. Liu, W. Guo, and M. A. Gebbie, 'Tuning Ionic Screening To Accelerate Electrochemical CO2 Reduction in Ionic Liquid Electrolytes,' *ACS Catalysis,* vol. 12, no. 15, pp. 9706-9716, 2022.

393c Origin of Copper Dissolution in Electrocatalytic Reduction Conditions Involving Amines

Yani Guan, Philippe Sautet

Cu dissolution has been identified as the dominant process that causes cathode degradation and losses even under cathodic conditions involving methylamine. Despite extensive experimental research, our fundamental and theoretical understanding of the atomic-scale mechanism for Cu dissolution in electrochemical conditions, eventually coupled with surface restructuring processes, is limited. Here, driven by the observation that the working Cu electrode is corroded using mixtures of acetone and methylamine even under reductive potential conditions (-0.75V *vs* RHE), we employed Grand Canonical density functional theory to understand this dynamic process under potential from a microscopic perspective. We show that amine ligands in solution directly adsorb on the electrode, coordinate with the metal center, and drive the rearrangement of the copper surface by extracting Cu as adatoms in low coordination positions, where other amine ligands can coordinate and stabilize a surface copper-ligand complex, finally forming a detached Cu-tetramine dication complex in solution, even at negative potential conditions. Calculations predict that dissolution would occur for a potential of -0.7 V *vs* RHE or above. Our work provides a fundamental understanding of Cu dissolution facilitated by surface restructuring in amine solutions under electroreduction conditions, which is required for the rational design of durable Cu-based cathodes for electrochemical amination or other amine involving reduction processes.

393d Advancing Boron Removal Performance in Capacitive Deionization

Betul Uralcan, Weiyi Pan, Menachem Elimelech, Amir Haji-Akbari

The increasing global demand for clean water in the face of freshwater scarcity has prompted a quest for innovative solutions. In this context, capacitive deionization (CDI) emerges as a promising desalination technology and an energy efficient alternative for selective ion removal particularly in scenarios involving low salt concentrations. Ion selectivity in CDI systems can be modulated through diverse mechanisms including modification of functional groups on nanoporous electrode materials. In this work, we use molecular simulations to explore the effect of surface functional group modifications on porous carbon electrodes to facilitate selective boron electrosorption. Our findings show that surface modification significantly enhances electrode performance for boron removal and selectivity. Specifically, we identify hydroxyl groups as a crucial factor in improving boron removal capacity and illustrate the underlying mechanisms of boron selectivity. The identification of molecular features for enhancing boron removal and selectivity demonstrates the role of interactions between functional groups and boron in regulating the electrosorption process.

393e Growth of Hexagonal Boron Nitride Frommolten Nickel Solutions: A Reactive Moleculardynamics Study

Jeffrey Comer, Amin Ahmadisharaf

Hexagonal boron nitride (hBN), is recognized for its exceptional attributes suitable for applications in electronics, catalysis, and nanosensor technology. Despite the potential, the atomic mechanisms underlying hBN formation, particularly through metal flux methods, still need to be explored. This study employs classical reactive molecular dynamics simulations to demystify the synthesis dynamics of hBN from liquid nickel, a process pivotal for high-quality hBN crystal formation. Our investigations reveal that hBN formation predominantly occurs at the liquid nickel surface, initiated by the interaction between N2 and nickel-solvated boron atoms. This interaction facilitates the emergence of intermediate species crucial for hBN nucleation. The study underscores the low solubility of N2 in bulk nickel and the consequential surface segregation of boron as critical factors influencing hBN synthesis. Furthermore, our findings highlight a temperature-dependent formation mechanism, offering insights into optimizing synthesis conditions for enhanced hBN quality and yield.
Our analysis elucidates the synthesis pathways, from the segregation of boron to the liquid nickel surface through to the formation of 1hBN, thereby providing a foundational understanding of the process. This study contributes to our understanding of hBN synthesis and provides opportunities for improved production techniques that will meet the increasing need for hBN in a range of high-tech applications. Chemical engineering benefits greatly from the novel approach we used and the important findings we produced, especially when creating cutting-edge materials.

393f Justification of a Pseudo-Initial State in Proton Transfer during the Volmer Reaction

Juye Kim, Andrew A. Peterson

Protons play a vital role in electrochemical reactions, particularly in the Volmer reaction, the most basic electrochemical process. This reaction involves the adsorption of a hydrogen atom on a surface by combining a proton with electrons. To simulate the Volmer reaction, we first consider an initial scenario where a proton originates from a bulk reservoir while a vacant metal slab is present on the surface. Before assessing the reaction barrier, the proton needs to migrate from the reservoir to the vicinity of the electrode surface. This transition to the near-surface layer is termed the pseudo-initial state. A key inquiry arises regarding how the proton travels from the bulk to the near-surface and whether this pseudo-initial state is indeed observable. If the proton doesn't localize at the near-surface layer but instead follows a concerted mechanism, the existence of the pseudo-initial state becomes doubtful. In this research, we explored proton pathways within the grand canonical system using the solvated jellium approach. Additionally, we investigated proton localization near the surface, confirming the presence of pseudo-initial states.

393g Enhancing Reversibility in Adsorption Processes: The Hysteresis Challenge

Filip Formalik, Jesse Prelesnik, J. Ilja Siepmann, Randall Snurr

Adsorption hysteresis can significantly impact the functionality of porous materials like metal-organic frameworks (MOFs), influencing the reversibility of adsorption in applications such as adsorption cooling and water harvesting, where reversibility is essential. Hysteresis is especially pronounced in MOFs with large pores, which hinders the use of these materials despite their potential for high adsorption capacity. Adsorption hysteresis arises from the presence of high energy barriers separating low- and high-density phases of the fluid confined in the pore and is coupled with the tendency of the system to remain in metastable phases.

In this study, we explore adsorption hysteresis through the van der Waals (vdW) loop, a metric that quantifies the stability limits of adsorption hysteresis. The vdW loop maps out the density of stable, metastable, and unstable phases of the confined fluid, utilizing spinodal and binodal points to define the stability limits of the phases and the equilibrium pressure for coexisting phases, respectively. To accurately predict the vdW loop, we employ transition matrix Monte Carlo (TMMC) simulations. These simulations offer a comprehensive description of the free energy landscape associated with adsorption, enabling the capture of all possible system macrostates, including unstable regions of the phase diagram (e.g., barriers between states). Our analysis leverages the gRASPA simulation software, developed in our group and optimized for GPU nodes, to significantly accelerate the adsorption simulations.

Our findings illustrate the effects of the simulation system's size on the width of the hysteresis loop. These effects, in turn, correlate with the barrier associated with the transition between low- and high-density phases. We further explore how specific interaction patterns on the surfaces of idealized pore geometries (such as *hydrophobic* and *hydrophilic* sites) can influence these properties. Patterns explored in model systems can be correlated with the distribution of functional groups in real MOFs. This relationship is important for the design of MOFs optimized for minimal adsorption hysteresis and for enhancing their efficiency in applications like adsorption cooling and water harvesting. Additionally, we establish connections between the hysteresis loop, the transition barrier, and particular pore shapes and system sizes in MOFs, offering new insights into the design of new materials showing reversible adsorption.

393h Effects of Environmental Species on Tribological Properties of MoS_2: Using Simulations to Interpret Experimental Observations

Scott Bobbitt, John Curry, Tomas Babuska, Michael E. Chandross

Molybdenum disulfide (MoS_2) is a lamellar solid with applications in semiconductor devices, catalysis, and lubrication. The layers of MoS_2 interact via weak van der Waals forces, enabling the layers to slide with a low coefficient of friction in dry or vacuum environments. This makes MoS_2 an attractive solid lubricant for use in satellites and other aerospace applications. However, exposure to environmental species, e.g. oxygen and water, has been shown to degrade the lubricating properties of MoS_2. The precise mechanisms of the interactions between MoS_2 and water and how these interactions affect its tribological properties remain poorly understood.

In this talk, we will present our theory for how adsorbed water impacts the tribology of MoS_2 lubricants, based on experiments and simulations. We report the coefficient of friction over a wide range of humidities using vacuum-based tribological experiments and describe how different types of molecular simulation (density functional theory, Monte Carlo, and molecular dynamics) augment our interpretation of those experiments. We will present computed adsorption isotherms for water adsorption on MoS_2 under realistic atmospheric conditions We will discuss the interaction of water and oxygen with common defects in MoS2, including sulfur vacancies, edge defects, grain boundaries, and oxidized sites. Finally, we will discuss how water adsorption on MoS_2 impacts the tribological performance of the material.

SNL is managed and operated by NTESS under DOE NNSA contract DE-NA0003525

393i DFT and Machine Learning Investigations on the Sorptive Removal of Trace Contaminants from Syngas

Dwijraj Mhatre, Divesh Bhatia

Introduction

To utilize the syngas derived from the gasification of carbonaceous feedstocks for the production of valuable chemicals, the syngas has to be free from various toxic trace contaminants such as Hg, P Se, As, Cd, and Pb. These elements get released during gasification, incineration, or pyrolysis in their elemental form, hydride form (PH_3, H_2Se, AsH_3), or sulfide form (PbS). They need to be removed since they are potential poisons to the catalyst(s) employed downstream for syngas processing [1]. The commercially available carbon and sulfur-based adsorbents are inefficient for the removal of trace contaminants at high temperatures. In the present work, the sorption of various gas-phase contaminants, viz., Hg^0, PH_3, H_2Se, AsH_3, Cd, and Pb is studied on monometallic and bimetallic surfaces, and machine leaning models are developed to predict the adsorption energy of the contaminants on bimetallic surfaces. Insights are obtained into the dissociation of AsH_3, H_2Se, and PH_3 on metals and into the segregation behaviour of Hg, As, Se, P, Cd, and Pb in alloys.

Methodology

First-principles density functional theory (DFT) calculations are performed to calculate the adsorption energy and to evaluate the most stable geometric configurations for the adsorption of gas-phase Hg^0, PH_3, H_2Se, AsH_3, Cd, and Pb on various metals and bimetallic surfaces. The calculations are carried out using the Vienna ab initio simulation package (VASP) and by employing the Perdew-Burke-Ernzerhof (PBE), optPBE-vdW, and PBE-D3 functionals of the generalized gradient approximation. The reaction energy pathway for PH_3 dissociation on various metals is studied using the CI-NEB method. PH_3 dissociation on the (1 1 1) surface of various metals and Ru(0 0 0 1) surface is considered to occur by the sequence of the elementary reactions described by Eqs. (R1)-(R4), where the subscript 'g' represents the gas-phase species and the subscript 'ad' represents the adsorbed surface species.

$PH_{3\,(g)} \rightarrow PH_{3\,(ad)}$ (R1)

$PH_{3\,(ad)} \rightarrow PH_{2\,(ad)} + H_{(ad)}$ (R2)

$PH_{2\,(ad)} \rightarrow PH_{(ad)} + H_{(ad)}$ (R3)

$PH_{(ad)} \rightarrow P_{(ad)} + H_{(ad)}$ (R4)

Similarly, calculations are performed considering the step-wise abstraction of H atoms from H_2Se and AsH_3 as well, and the reaction energy barriers and the reaction energies for the elementary reactions are calculated.

Machine learning models are developed to predict the DFT-calculated adsorption energy of trace contaminants on various bimetallic surfaces using scikit-learn in Python [2]. Linear algorithms viz., ordinary linear regression (OLR), partial least-squares (PLS) regression, and least absolute shrinkage and selection operator (LASSO) regression, and non-linear tree ensemble algorithms viz., extra tree regression (ETR), random forest regression (RFR), and gradient boosting regression (GBR) are used. A set of 29 descriptors comprising of physical properties of the constituent metals in the adsorbents, electronic properties of the adsorbent surfaces, and a few other properties of the adsorbent surface are initially considered. From this initial set of descriptors, 17 descriptors are shortlisted on the basis of their noncollinearity.

Results and Discussion

The adsorption of Hg^0, Cd, and Pb is the most favourable on the hollow sites of most metal surfaces, whereas the adsorption of hydrides PH_3, H_2Se, and AsH_3 is favourable on the top sites of most metals considered in this study. The adsorption of gas-phase Hg, PH_3, H_2Se, AsH_3, Cd, and Pb is exothermic on various metals. Further, Rh, Ru, Pd, Pt, and Ir are found to be the most promising metals for the removal of these contaminants from gas streams. The adsorption energies of PH_x (x=1, 2, 3) on various metals scale with the adsorption energy of elemental P. The stepwise dissociation of PH_3 is evaluated on various metals by reactions R1-R4, and the reaction energy diagram is shown in Fig. 1. The analysis of the calculated reaction energies and activation barriers shows that PH_3 dissociation is more facile on Rh, Ru, Pd, Ir, and Pt as compared to Cu, Ag, and Au. Similar results are observed for the dissociation of H_2Se and AsH_3. The segregation energy for the elements present on the surface of various alloys is plotted in Fig. 2 (a). The negative values of segregation energy indicate the tendency of elements Hg, As, Se, P, Cd, and Pb to segregate towards the surface of Ru, Rh, Pt, Ir, Pd, Cu, Ag, and Au after alloy formation. The formation energy calculations indicate the possibility of the formation of alloys between metals Ag, Au, Cu, Ir, Pd, Pt, Rh, and Ru and elements Hg, As, Se, P, Cd, and Pb (Fig. 2 (b)).

DFT calculations are performed to evaluate the binding of the trace contaminants on 56 different compositions of bimetallic adsorbents. Bimetallic adsorbents with an overlayer of Ru, Rh, Ir, or Pt on Ag or Au exhibit a higher affinity for PH_3 as compared to the monometallic adsorbents (Fig. 3). Similar results are obtained for the adsorption of Hg, H_2Se, AsH_3, Cd, and Pb. The adsorption of these contaminants on the overlayered bimetallics is stronger than that on monometallic surfaces due to the ligand and strain effects.

Six ML algorithms are used to predict the adsorption energy of various trace contaminants on overlayered bimetallic surfaces using the 17 shortlisted descriptors as inputs. The tree ensemble methods, viz. ETR and RFR, result in the best prediction with a low value of prediction error ($RMSE_{test}$) (Fig. 4 (a)). Moreover, the standard deviation for $RMSE_{test}$ is quite low for the tree ensemble models. The significance of each of the 17 shortlisted descriptors in predicting the adsorption energy of trace contaminants is investigated using the feature importance score in the tree-based ETR algorithm, and the top three descriptors are evaluated to be the upper edge of the d-band ( ), group number of the constituent metal present in the overlayered bimetallic surface, and d-band center ( ). Further, the average $RMSE_{test}$ using the ETR model increases marginally (from 16.2 to 17.4 kJ/mol for AsH_3) when the number of descriptors is reduced from 17 to 3 (Fig. 4 (b)). Thus, the ETR model can be used for an accurate prediction of the adsorption energy of AsH_3 and other trace contaminants using only three input parameters.

Significance of the work

Metals and novel bimetallic adsorbents are identified for the removal of various trace contaminants, and the energetics involved in the dissociation of PH_3, H_2Se, and AsH_3 on metals is shown. Tree-ensemble-based machine learning models are developed to accurately predict the adsorption energy of various trace contaminants on bimetallic adsorbents.

References

[1] C.J. Howard, R.A. Dagle, V.M. Lebarbier, J.E. Rainbolt, L. Li, D.L. King, Progress toward biomass and coal-derived syngas warm cleanup: Proof-of-concept process demonstration of multicontaminant removal for biomass application, Ind. Eng. Chem. Res. 52 (2013) 8125–8138.

[2] F. Pedregosa, G. Varoquaux, A. Gramfort, V. Michel, B. Thirion, O. Grisel, M. Blondel, P. Prettenhofer, R. Weiss, V. Dubourg, J. Vanderplas, A. Passos, D. Cournapeau, M. Brucher, M. Perrot, É.

Duchesnay, Scikit-learn: machine learning in python, J. of Machine Learn. Res. 12 (2011) 2825–2830.

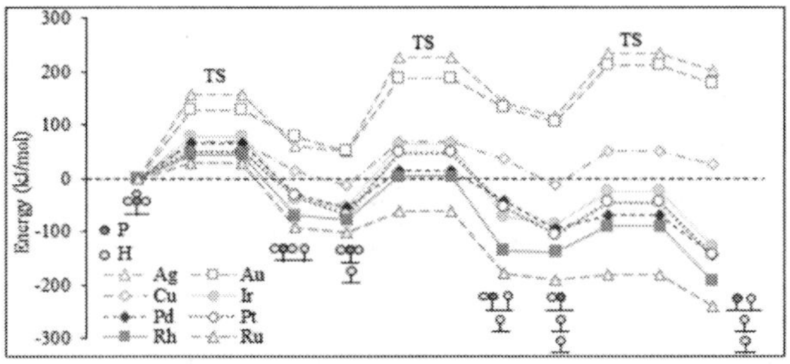

Fig. 1. Energy diagram for PH_3 dissociation on (1 1 1) surface of various metals and Ru(0 0 0 1).

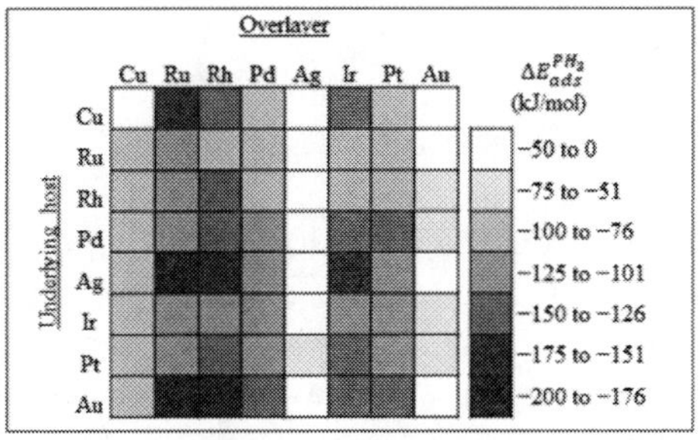

Fig. 3. PH_3 adsorption energy ($\Delta E_{ads}^{PH_3}$) over various overlayered bimetallic surfaces

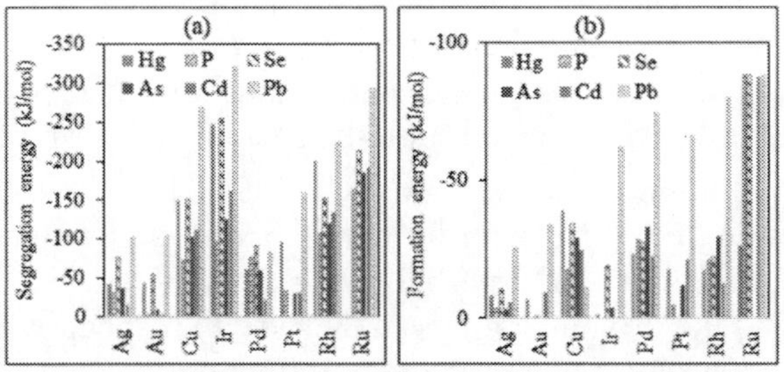

Fig. 2. (a) Segregation energy and (b) formation energy of various alloys with trace elements present in the top-most layer of the surface.

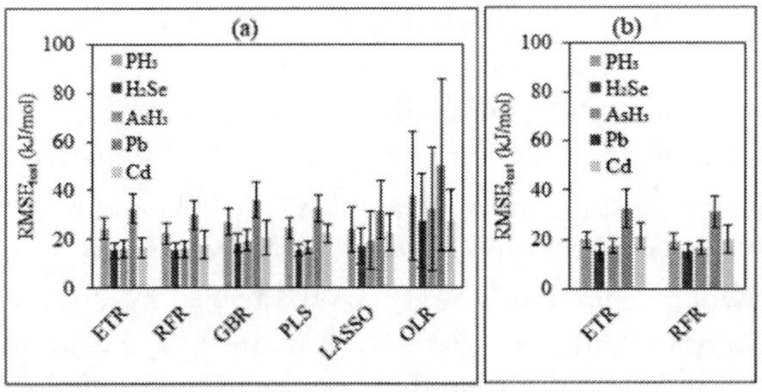

Fig. 4. Average RMSE for predicting adsorption energy of trace contaminants using different ML models employing (a) 17 descriptors and (b) top 3 descriptors.

393j Polyoxometalate-Based Membranes for Water Purification: A Molecular Dynamics Study

Vinay Thakur, Raghavan Ranganathan

Rising population and pollution levels have spurred an increased need for access to clean water across the globe. The inadequate access to clean water has led to a high risk of tumors, congenital disorders, and annual economic losses. Polyoxometalate-based porous material shows great promise as filtration membranes for removing heavy metal ions and other toxic and hazardous chemicals from the contaminated water Polyoxometalates are anionic nanoclusters composed of oxo ligands and early transition metals (V, Mo, W or Ta) in their highest oxidation state. Besides, they have shown a wide range of applications in the field of medicine, catalytic converters, electromagnetism, and energy storage materials. Motivated by the excellent efficacy demonstrated by polyoxometalate membranes in size-based separation of cationic, anionic and neutral dyes by our experimental collaborators, in this study, we use non-equilibrium classical molecular dynamics to study the permeation of water and industrial dyes through the system composed of polyoxometalate anion $[P8W48O184]^{40-}$ and tetraheptyl ammonium counterions (C7H15). The polyoxometalate membrane is held fixed to allow greater degree of freedom for contaminants and water molecules. The study involves the use of ESP(CHELPG) and AM1-BCC partial charges generated separately using DFT calculations for polyoxometalate and the organic entities in the system respectively. The non-bonded interactions between the membrane, counterions, water molecules and model contaminants are modeled using the Lennard-Jones potential and Coulombic interactions. Two different configurations of the polyoxometalate clusters are considered, namely, ordered (periodic arrangement of pores with pore features modeled to reproduce experimental observations) and disordered (where pores are placed randomly). The study elucidates mechanisms of water permeation through the polyoxometalate membrane at different pressure ranges and calculates the variation in water flux as a function of pressure. Furthermore, rejection of toxic contaminants such as common organic dyes are studied, and the removal efficiency are correlated with the pore and contaminant size and the nature of interactions with the polyoxometalate membrane. Detailed structural (via radial distribution functions) and transport (via diffusivities) properties of the contaminant and water are analyzed and the efficacy of polyoxometalate membranes in water treatment are established.

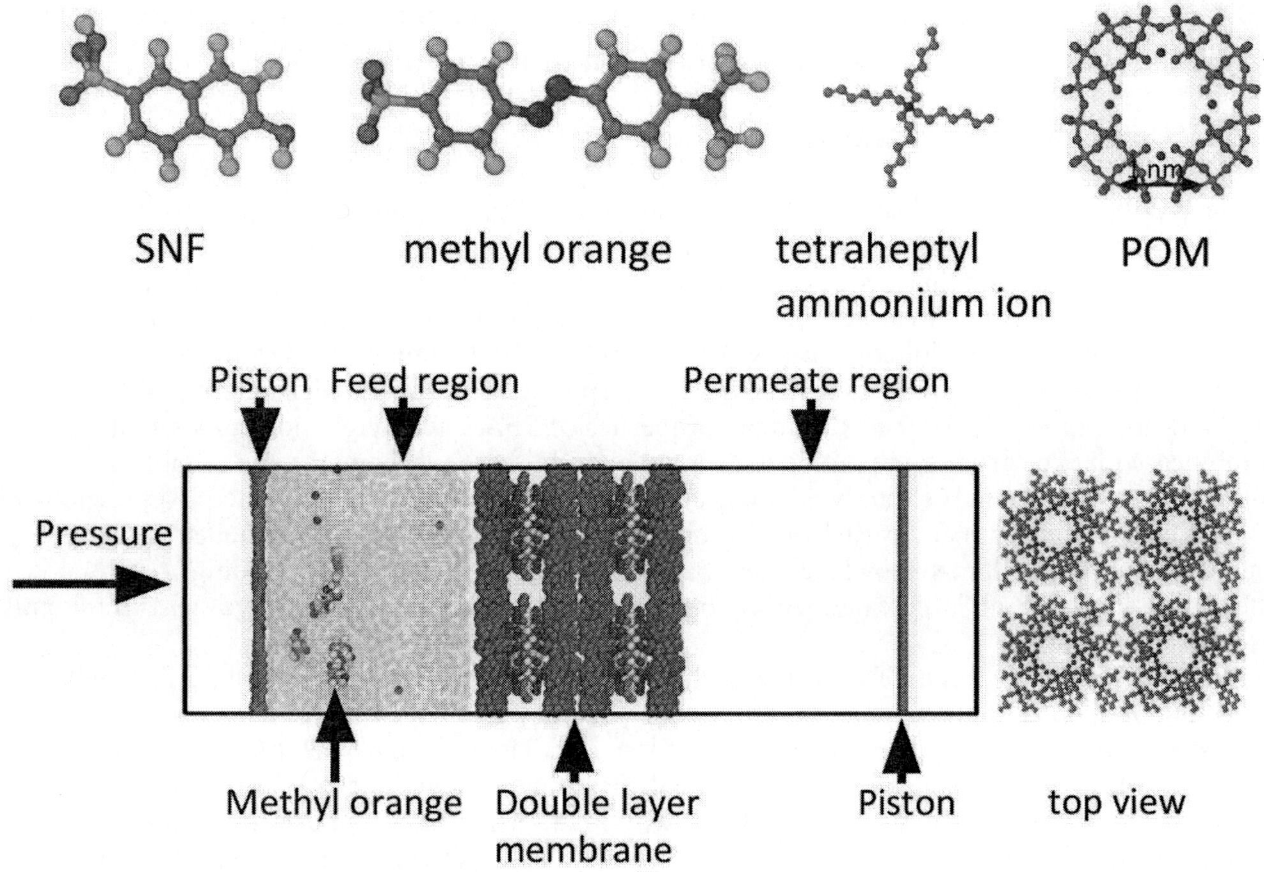
SNF
methyl orange
tetraheptyl ammonium ion
POM
1 nm
Piston
Feed region
Permeate region
Pressure
Methyl orange
Double layer membrane
Piston
top view

393k An Analysis of Fluorinated and Fluorine-Free Surfactants at Interfaces: Insights from Ab Initio and Molecular Dynamics Simulations.

Gabriel D. Barbosa, Alberto Striolo

Per- and poly-fluoroalkyl substances (PFAS) constitute a class of synthetic compounds with exceptional interfacial properties. Their widespread use in many industrial applications and consumer products, combined with their remarkable chemical and thermal stability, has led to their ubiquitous presence in environmental matrices, including surface and groundwater. Mounting evidence shows their accumulation in the biological tissues of both humans and animals. To design better fluorine-free chemical alternatives, it is necessary to develop a deep molecular-level understanding of the mechanisms responsible for the exceptional properties of PFAS. Although molecular simulations, combined with experiments, are promising for uncovering these mechanisms, the reliability of simulation results depends strongly on the accuracy of the force fields implemented. At the moment, atomistic force fields are not available to describe PFAS in a variety of environments. Ab initio methods could help fill this knowledge gap, but they are computationally demanding. As an alternative approach, ab initio calculations could be used to develop accurate force fields for atomistic simulations.

In this work, atomistic force fields were developed for perfluorinated sulfonic and perfluoroalkyl acids, building on ab initio simulation results. The accuracy of the new force fields was benchmarked against solvation-free energy and interfacial tension data. The new force fields were then used to probe the interfacial behavior of PFAS surfactants. Both thermodynamic and structural interfacial properties were calculated. To help understand why PFAS sometimes outperforms non-fluorinated chemicals, the PFAS interfacial properties observed in our simulations were compared against those manifested by different fluorine-free surfactants. The good agreement achieved with experiments suggests that the proposed protocol could be implemented to study other perfluorinated substances.

673a High-Throughput Virtual Screening and Hybrid Computational/Experimental Active Learning for Novel Hydrogel-Nanoparticle Sensors for Fentanyl Detection

Armin Shayesteh Zadeh, Alexander Winton, Matthew Coppock, Joseph M. Palomba, Andrew Ferguson

Fast, portable, and reliable detection of toxic chemical and biological compounds is an important challenge in many security and healthcare applications. Chemical sensors based on photonic crystals that use protein-catalyzed capture (PCC) agents as receptors are promising tools for on-the-fly detection of chemical and biological compounds. We report the development and deployment of a high-throughput screen utilizing experimental and computational data to discover PCC agent candidates that maximize binding affinity and specificity for fentanyl detection. We employ a multi-fidelity, multi-objective Bayesian optimization (MFMOBO) approach capable of asynchronous ingestion of computational predictions and experimental measurements to efficiently search the PCC design space for candidates that maximize fentanyl binding affinity and binding specificity relative to a decoy molecule. The MFMOBO approach performs guided sampling of a small fraction of chemical space to efficiently resolve the affinity-specificity Pareto frontier and identify the top-performing PCC candidates for incorporation into next-generation chemical sensing devices.

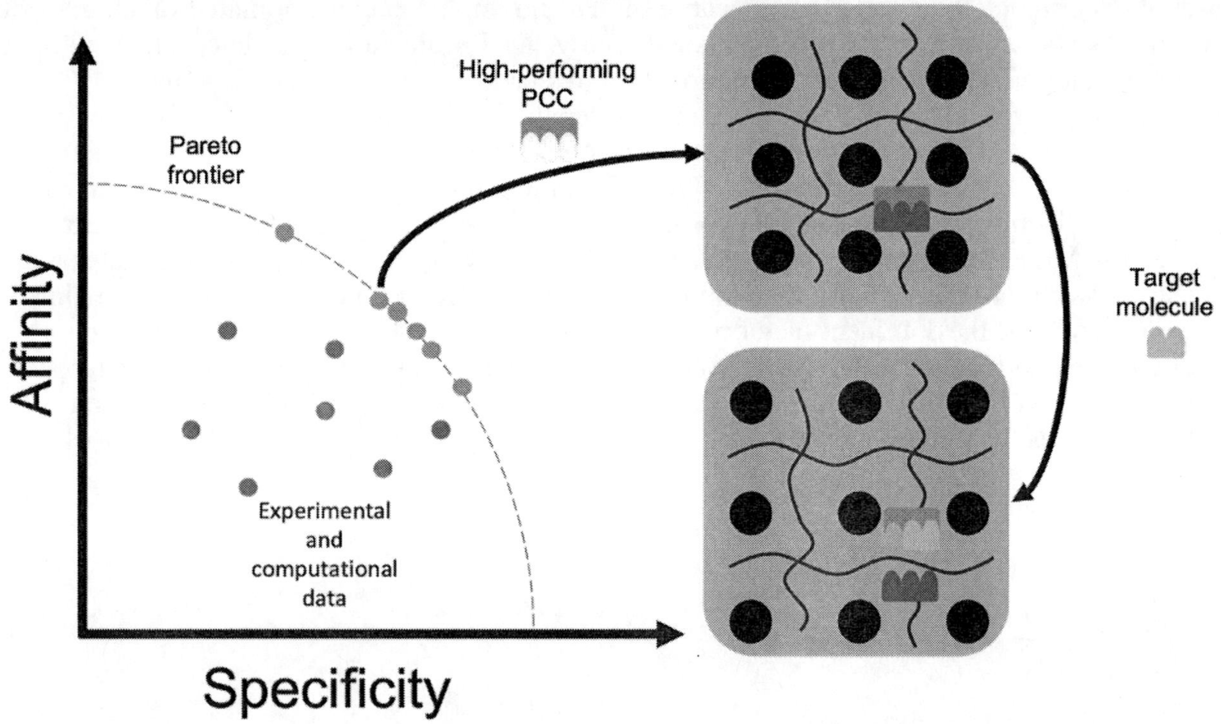

673b Computational Discovery of Metal-Organic Frameworks with High Water Uptake Capacity for Next-Generation Membranes

Akash Ball, Shuwen Yue, Gianmarco Terrones, Heather Kulik

Developing next-generation membranes for ion-selective separation is paramount for applications like desalination, battery recycling, and heavy metal recovery. Still, traditional membranes often have high permeability-selectivity tradeoffs, slow water permeation, complex fabrication routes, and defects. Recent studies have demonstrated that metal-organic frameworks (MOFs) can have more than one order of magnitude higher water flux than current state-of-the-art 2D membranes like graphene and retain almost 100% ion rejection rate. Although some qualitative structure-property relationships exist in the water uptake capacity of MOFs, a quantitative understanding of how MOF linker chemistry and pore geometry affect water uptake capacity is still unknown. Moreover, the diversity of MOF design space is in the order of millions, making it impossible to identify MOFs with high water uptake capacity by performing experiments and computer simulations. Here, we take a computational approach to accelerate discovery of design rules for MOFs capable of high water uptake. We choose a subset of experimentally reported water-stable MOFs and implement an automated functionalization procedure. Next, we perform density functional theory (DFT) calculations to determine the binding affinity of water with a diverse set of functional groups present on the linkers from these MOFs The DFT calculations reveal that the functional groups capable of hydrogen bonding with water molecules result in the strongest binding affinity. Next, we calculate the water uptake capacity of functionalized MOFs from grand canonical Monte Carlo (GCMC) simulations. From the GCMC simulations, we discover that electronegative and small functional groups significantly improve the water uptake capacity of a MOF. Furthermore, we find that the water uptake capacity of MOFs correlates much more strongly with MOF pore size than the water-functional group binding affinity, suggesting MOF pore geometry to be the dominant factor in water uptake capacity. Finally, we develop machine learning models to identify water-stable MOFs in experimental and hypothetical MOF databases by performing virtual high-throughput screening and suggesting design guidelines for selecting functional groups that improve water uptake capacity without compromising the water stability of MOFs.

673c Bespoke Nanoparticle Synthesis and Chemical Knowledge Discovery Via Autonomous Experimentations

Hyuk Jun Yoo, Nayeon Kim, Heeseung Lee, Daeho Kim, Leslie C. O. Tiong, Hyobin Nam, Chansoo Kim, Seung Yong Lee, Kwan-Young Lee, Donghun Kim, Sang Soo Han

In this study, we present an innovative approach for optimizing the synthesis of metal nanoparticles (NPs) with specific optical properties. Traditional methods of exploring various synthetic variables are time-consuming and costly. Our solution involves an autonomous experimentation platform that integrates a batch synthesis module for metal NPs with a UV-Vis spectroscopy module, guided by AI optimization modeling. Using silver (Ag) NPs as a model system, we demonstrate the effectiveness of our approach, achieving precise control over the desired absorption spectra within a remarkably low number of iterations (200 iterations when optimizing among five different aqueous synthetic reagents) at room temperature. Furthermore, our analysis of synthetic variables uncovers a novel chemistry pertaining to the role of citrate in Ag NP synthesis. We find that the quantity of citrate plays a crucial role in modulating the competition between spherical and plate-shaped NPs, consequently influencing the shapes of the absorption spectra. Overall, our study showcases the platform's ability to expedite search processes and uncover new chemical insights through the analysis of data gathered from autonomous experiments.

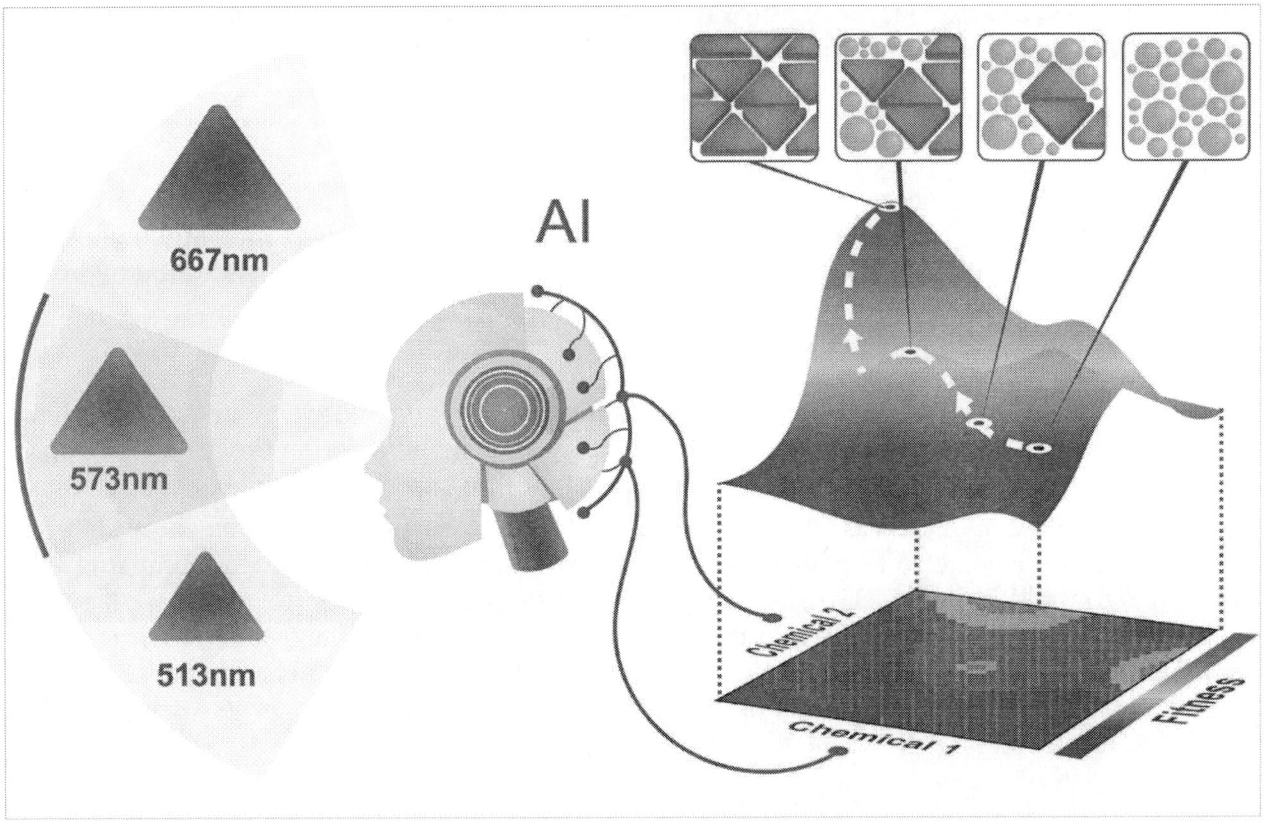

673d Multi-Objective Bayesian Optimisation for Sustainable Reactions Development

Jiyizhe Zhang, Alexei A. Lapkin

Optimising chemical reactions to discover the best conditions is important at multiple stages of reaction development, from lab-scale condition screening to large-scale manufacturing. Often multiple objectives need to be considered alongside yield optimisation, for example, achieving high yield with high selectivity, considering economic and sustainability aspects at the same time, or optimizing yield and throughput simultaneously for scale-up. Finding such optimal solutions usually requires searching in a large design space for a set of Pareto solutions rather than one solution, which tends to be tedious and time-consuming. In this study, we focus on optimising two chemical reactions in flow with multiple objectives using Bayesian optimisation, to consider space-time-yield and sustainability factor at the same time.

Amide bond formation is one of the most prevalent reactions in pharmaceutical industry, among which the Schotten-Baumann reaction with a long history is useful as a potential green amide formation approach. However, the use of water in the reaction system often causes undesired hydrolysis and can generate a multiphase system. This makes the reaction space complex and challenging to find the optimal conditions. In the first case study, a Schotten-Baumann reaction was studied in continuous flow and was optimised with two objectives using a Bayesian optimisation algorithm based on the *q*-Noisy Expected Hypervolume Improvement (*q*NEHVI)[1]. The algorithm guided the experiment design over a range of mixed variables, i.e. electrophiles, equivalents, solvents, and flow rates, and was able to identify the Pareto front of optimal solutions efficiently. Based on the optimisation results, reaction under flow and batch conditions were compared; undesired hydrolysis was suppressed successfully using the flow conditions. Finally, the relationship between solvent and flow rate was discussed to gain more insights into this reaction.

In the second case study, we focus on robust applications where large noises are unavoidable. This is useful when dealing with unstable reagents and for large-scale manufacturing processes. To cope with large noises, a multi-objective Bayesian optimisation algorithm based on Euclidian expected quantile improvement (MO-E-EQI[2]) was adopted to consider heteroscedastic noises. First, an *in silico* benchmarking was conducted to evaluate algorithm performances. MO-E-EQI was compared to two state-of-the-art multi-objective Bayesian algorithms and showed the best performance for a noise level up to 20%. Then MO-E-EQI was implemented in a real case to optimize an esterification reaction in flow with four continuous variables: flow rates, equivalents, catalyst loadings, temperatures. The algorithm efficiently identified a clear trade-off under noises between space-time yield and E-factor.

In conclusion, the study demonstrates multi-objective reaction optimisation using Bayesian algorithms for two complex systems: multi-phase reaction system and noisy system. A set of Pareto front solutions can be found efficiently guided by Bayesian approach, which shows a promising way to allow multiple factors to be considered at early-stage reaction development.

673e Improving Uncertainty Estimation Based on Latent Space Distance for Graph Neural Network Potentials

Joseph Musielewicz, John Kitchin

Graph neural networks (GNNs) have been shown to be astonishingly capable models for molecular property prediction, particularly as surrogates for expensive density functional theory calculations of relaxed energy for novel material discovery. However, one limitation of GNNs in this context is the lack of useful uncertainty prediction methods, as this is critical to the material discovery pipeline. In this work, we show that uncertainty quantification for relaxed energy calculations is more complex than uncertainty quantification for other kinds of molecular property prediction, due to the effect that structure optimizations have on the error distribution. We propose that distribution-free techniques are more useful tools for assessing calibration, recalibrating, and developing uncertainty prediction methods for GNNs performing relaxed energy calculations. We also develop a relaxed energy task for evaluating uncertainty methods for equivariant GNNs, based on distribution-free recalibration and using the Open Catalyst Project dataset. We benchmark a set of popular uncertainty prediction methods on this task, and show that latent space distance methods, with our novel improvements, are the most well- calibrated and economical approach for relaxed energy calculations.

673f Multi-Objective Inverse Molecular Design of CO2 Capturing Agents

Zisheng Zhang, Jenny Yang, Anastassia Alexandrova

Aqueous direct air capture (DAC) is a key technology toward a carbon negative infrastructure. Developing sorbent molecules with water- and oxygen-tolerance and high CO2 binding capacity is therefore highly desired.

Herein, we analyze the CO2 absorption chemistries on amines, alkoxides, and phenoxides with density functional theory calculations, and perform inverse molecular design of the optimal sorbent. The alkoxides and phenoxides are found to be more suitable for aqueous DAC than amines thanks to their water-tolerance (overall lower pKa makes alkoxides less prone to protonation by water) and capture stoichiometry of 1:1 (2:1 for amines). All three molecular systems are found to obey the same linear scaling relationship (LSR) between pK_(CO_2) and pK_a, since both CO2 and proton are bonded to a nucleophilic site (alkoxy or amine) through a σ-type bonding orbital. Several high-performance alkoxides are proposed from the computational screening. In contrast, phenoxides have relatively poor correlation between pK_(CO_2) and pK_a, showing promise for optimization.

We apply genetic algorithm to search the chemical space of substituted phenoxides for the optimal sorbent that 'escape' the LSR. The most promising off-LSR candidate phenoxides feature bulky ortho substituents forcing the CO2 adduct into a perpendicular configuration with respect to the aromatic ring. In this configuration, CO2 and a proton from water bind to the phenoxide using different molecular orbitals, thereby decoupling the pK_(CO_2) and pK_a. The pK_(CO_2)-pK_a trend and off-LSR behaviors are then confirmed by experiments, validating the inverse molecular design framework. This work not only extensively studies the chemistry of the aqueous DAC, but also presents a transferrable computational workflow for understanding and optimization of other functional molecules.

673g Computational Protein-Ligand Co-Design for Enhancing Extracellular Electron Transfer in *Lactiplantibacillus Plantarum*

Vladimir Porokhin, Soha Hassoun, Anne Brown

While directed evolution (DE) of proteins has been tremendously successful in improving protein selectivity, specificity, expression, solubility, and stability, goal-oriented protein design remains a challenging problem. As the number of possible protein mutants is large, even when limiting potential mutation to a small number of residues, a deliberate design strategy is required. Rational design [1, 2] and DE are two useful experimental options that are now complemented by machine learning (ML) techniques that elucidate the protein-function landscape by computational means [3]. An equally important and challenging problem is engineering of ligands with specific structural motifs or chemical properties. Similarly to protein engineering, the space of small molecules is large, likely in excess of 10^{60} structures. Generative deep learning models, either conditioned on properties of interest, or guided by optimization algorithms have been proposed as a promising strategy for identifying molecules with desired characteristics [4].

The prior approaches subscribe to the *individual design* paradigm, where proteins and small molecules are engineered individually. This paradigm is sufficient in many endeavors where the goal is to engineer an entity that will function within an existing system (e.g., a drug targeting a particular receptor, or an enzyme to increase the yield for a target molecule). However, emerging applications in bioelectronic sensing [5] and synthetic biosynthesis [6] are not necessarily bound by the constraints of existing systems and thus may benefit from the *mutual co-design* of ligands and proteins, where both are engineered together. Our hypothesis is that co-design offers greater engineering flexibility, and thus a higher likelihood of discovering highly functional protein-ligand combinations, compared to individual design alternatives. Unfortunately, there are currently no methods that allow effective exploration of this joint protein-ligand design space.

We explore this co-design paradigm by considering type II NADH:quinone oxidoreductase (Ndh2) and its interaction with quinones to power extracellular electron transfer (EET). EET has been previously harnessed to sense various chemicals, thereby enabling creation of inexpensive whole-cell bioelectronic sensors [7]. Ndh2 is a ubiquitous protein that exists in many organisms as a key component of the respiratory chain, but was also found to facilitate EET in *L. plantarum* when in the presence of exogenous quinones [8]. As the EET response is dependent on the type of quinone and other factors, Ndh2 presents an exciting opportunity for designing sensors for pharmacologically relevant quinones [7]. The complex dependency of Ndh2 variants and quinones underscores the need to consider the Ndh2 and the quinone in tandem if the objective is to design for maximum EET rate.

In this work, we develop a framework for exploring the co-design landscape of proteins and ligands and present its application to enhancing EET transfer when considering Ndh2-quinone interactions. Our method for co-design is composed of three major steps. First, we design a process for creating separate libraries of protein mutants and quinone molecule derivatives. The Ndh2 variants were constructed by enumerating all single-point mutations around the quinone binding site, while the ligands were made by iteratively adding functional groups, e.g., phenyl, hydroxy, and other groups, in up to 6 locations on 1,4-dihydroxy-2-naphthoic acid and menadione, two quinones with a significant effect on EET. Second, we implement two search strategies for exploring protein-ligand combinations. In individual design, either the protein or the ligand is allowed to vary, so we construct those protein-ligand pairs by selecting a protein or ligand from our libraries and combining it with a wildtype Ndh2 or one of the base quinones. In co-design, both the protein and ligand can vary, so we create those pairs by combining a protein variant and a derivative quinone sampled from the libraries. Finally, we develop a protocol for evaluating the goodness of fit between a quinone and Ndh2 using molecular docking with AutoDock Vina and predicted free energy of binding calculations with MM/GBSA. Our proposed protocol is designed to approximate the favorability of the Ndh2-quinone interaction, which we believe is a major contributor to EET activity, and we use it to identify protein-ligand pairs with greatest interaction favorability.

We report several findings. First, we demonstrate that our proposed co-design approach can consider a significantly larger number of protein-ligand pairs than protein- or ligand-centric individual design methods alone. Second, we show that the co-design approach identifies pairs with the highest interaction favorabilities, making it the superior alternative to traditional individual design. Finally, we present a number of promising Ndh2-quinone combinations and highlight their key features that we believe would help improve EET. By simultaneously considering both the ligand and the protein, we are able to design novel biochemical systems in ways that were not possible with prior methods.

References

1. Hellinga, H.W. Rational protein design: Combining theory and experiment. *Proceedings of the National Academy of Sciences*. 1997, 94, 19. doi: 10.1073/pnas.94.19.10015

2. Korendovych, I.V. Rational and Semirational Protein Design. *Protein Engineering, Methods in Molecular Biology*. 2018, 1685, 15-23. doi: 10.1007/978-1-4939-7366-8_2

3. Wang, Y., Xue, P., Cao, M., Yu, T., Lane, S.T., Zhao, H. Directed Evolution: Methodologies and Applications. *Chem Rev*. 2021;121(20):12384-12444. doi: 10.1021/acs.chemrev.1c00260

4. Gao, W., Coley, C.W. The Synthesizability of Molecules Proposed by Generative Models. *J. Chem. Inf. Model.* 2020, 60, 12, 5714-5723. doi: 10.1021/acs.jcim.0c00174

5. Shanbhag, M.M., Manasa, G., Mascarenhas, R.J., Mondal, K., Shetti, N.P. Fundamentals of bio-electrochemical sensing. *Chemical Engineering Journal Advances*. 2023, 16. doi: 10.1016/j.ceja.2023.100516

6. Bowie, J.U., Sherkhanov, S., Korman, T.P., Valliere, M.A., Opgenorth, P.H., Liu, H. Synthetic Biochemistry: The Bio-inspired Cell-Free Approach to Commodity Chemical Production. *Trends in Biotechnology*. 2020, 38, 7, 766-778. doi: 10.1016/j.tibtech.2019.12.024

7. Li, S., De Groote Tavares, C., Tolar, J.G., Ajo-Franklin, C.M. Selective bioelectronic sensing of pharmacologically relevant quinones using extracellular electron transfer in *Lactiplantibacillus plantarum*. *Biosens Bioelectron.* 2024 Jan 1:243:115762. doi: 10.1016/j.bios.2023.115762

8. Stevens, E.T., Van Beeck, W., Blackburn, B., Tejedor-Sanz, S., Rasmussen, A.R.M., Mevers, E., Ajo-Franklin, C.M., Marco, M.L. *Lactiplantibacillus plantarum* uses ecologically relevant, exogenous quinones for extracellular electron transfer. *bioRxiv* 2023.03.13.532228. doi: 10.1101/2023.03.13.532228

455f Machine Learning-Enabled Exploration of the Electrochemical Stability of Real-Scale Metallic Nanoparticles

Sang Soo Han

Surface Pourbaix diagrams are critical to understanding the stability of nanomaterials in electrochemical environments. Their construction based on density functional theory (DFT) is, however, prohibitively expensive for real-scale systems, such as several nanometer-size nanoparticles (NPs) involving at least thousands of noble metal atoms, and this limitation calls for machine learning (ML)-driven approaches. Herein, with the aim of accelerating the accurate prediction of adsorption energies for a wide range of surface coverages on large-size NPs, we developed a bond-type embedded crystal graph convolutional neural network (BE-CGCNN) model in which four bonding types were treated differently. Owing to the much enhanced accuracy of the bond-type embedding approach compared to the original CGCNN, we demonstrate the construction of reliable Pourbaix diagrams for very large-size NPs involving up to 6,525 atoms (approximately 4.8 nm in diameter), which enables the exploration of electrochemical stability over various NP sizes and shapes. We reveal that ML-based Pourbaix diagrams well reproduce the experimental observations with increasing NP size, such as the increasing O- to OH-covered phase ratio and the decreasing Pt dissolution phase in the diagrams. This work suggests a new method for accelerated Pourbaix diagram construction for real-scale and arbitrarily shaped NPs, which would significantly open up an avenue for electrochemical stability studies.

394a Polymerized Amyloids: Sequence and Structure Space of Tandem Repeat Proteins

Tianyi Jin, Connor Coley, Alfredo Alexander-Katz

Amyloids are highly ordered cross-β protein aggregates that have traditionally been associated with neurodegenerative diseases. Recent research underscores their potential in the realm of functional materials based on their unique repetitive structures, high stability, and self-assembling capabilities. Leveraging advancements in protein structural prediction tools, our study explores tandem sequence designs to achieve varied aggregation morphologies with high in silico structural confidence, facilitated by machine learning-driven evolution. We present a robust approach that promotes amyloid β-42 aggregation into β-solenoid-like structures through the polymerization of amyloid segments with a random peptide linker. Employing the AlphaFold model allows us to screen linker sequences efficiently, ensuring high structural confidence. Our methodology extends to rapidly explore the tandem sequence space using a genetic algorithm and develop a surrogate model of sequences-to-structural confidence predictions based on large language model. This approach not only yields β-solenoid structures but also enables the design of single-layer β-sheets along the evolutionary pathway. Our work provides new insights on amyloid aggregation, functional amyloid-based materials and protein scaffold designs.

394b Uncovering Interaction Pathways in Protein Dynamics Using Machine Learning Models

Vignesh Bhethanabotla, William Goddard III

In machine learning approaches to protein characterization and design, representations typically focus on either sequence-based or structure-based information. However, these representations may overlook the dynamic nature of proteins, crucial for their properties and functionalities. Analysis of molecular dynamics simulations offers insights into such dynamic interactions and the effect of minimal modifications to protein sequences and structures, including effects on conformational entropy and long-range allosteric effects, often missed by traditional sequence and structure-based models. These interactions can be particularly relevant in engineering strategies like directed evolution, a strategy of interest in the design of enzymatic systems. [1] Machine-learning models provide an avenue to analyzing rich dynamics data which, even without knowledge of the underlying equations of motion, can uncover directed relational interactions between the variables under study. [2]

A particularly successful strategy to engineering proteins, directed evolution, involves navigating sequence space from a starting point to optimize enzyme properties. This process can be guided randomly or by targeting specific sites for mutation. However, traditional iterative optimization approaches encounter challenges when dealing with epistatic interactions, where mutations' effects depend on higher-order interactions, such as modulation by ligands, substrates, allostery, or conformational dynamics. [3] Understanding these interactions can inform the selection of sites for directed evolution libraries (sites selected based on predicted importance to modifying the protein system's desired property or function), improving the efficiency of enzyme design strategies. We focus our system of interest on the beta subunit of tryptophan synthase, an enzyme responsible for the conversion of amino acid serine to tryptophan. In nature, this system is bound to a partner protein and exhibits an allosteric activation for its functionality, which can be recapitulated using directed evolution to promote non-natural, standalone activity. [4] Using a deep-learning approach to analyze data from molecular dynamics simulations of this system, we generate directed graph representations which encode dynamic, pair-wise interactions between residues in the wild-type and variants of the enzymatic protein, guiding choices of site targets. Comparisons of these results with traditional methods used in an experimental setting are also made and suggest potential extensions for this approach to computationally-assisted library design in other systems.

1. Romero, P. A.; Arnold, F. H. Exploring Protein Fitness Landscapes by Directed Evolution. Nat Rev Mol Cell Biol 2009, 10 (12), 866–876.

2. Löwe, S.; Madras, D.; Zemel, R.; Welling, M. Amortized Causal Discovery: Learning to Infer Causal Graphs from Time-Series Data. arXiv February 21, 2022.

3. Starr, T. N.; Thornton, J. W. Epistasis in Protein Evolution. Protein Science 2016, 25 (7), 1204–1218.

4. Buller, A. R.; Brinkmann-Chen, S.; Romney, D. K.; Herger, M.; Murciano-Calles, J.; Arnold, F. H. Directed Evolution of the Tryptophan Synthase β-Subunit for Stand-Alone Function Recapitulates Allosteric Activation. PNAS 2015.

394c Experimentally-Informed Optimization of Amino Acid Interactions for Predictions of Intrinsically Disordered Protein Phase Behavior

Beth Wei, Shiv Rekhi, Young C. Kim, Jeetain Mittal

The demixing of intrinsically disordered proteins (IDPs) into coexisting protein-deficient and protein-enriched phases through liquid-liquid phase separation (LLPS) underlies the formation of membraneless organelles (MLOs). The formation and dissociation of MLOs, which play a key role in numerous cellular processes such as transcription and stress response, displays sensitivity to the IDP sequence and environmental conditions. Due to the large length and time scales involved in the process of LLPS, coarse-grained (CG) models have been instrumental in studying the effects of the primary sequence of IDPs on their phase behavior. However, to accurately model LLPS of a diverse set of IDPs using CG models, two main challenges remain. First, an appropriate mapping of the desired system into the reduced representation is required. Second, the reduced representation must be transferable, i.e. the model needs to be able to describe most proteins. To counter these challenges, amino-acid specific interaction potentials are usually developed and refined against experimentally derived single-chain dimensions of a subset of IDPs and validated against in-vitro phase separation assays targeting a handful of residue substitutions. In this work, we examine whether more accurate models of protein LLPS can be developed by accounting for in-vitro phase behavior in the optimization of interaction potentials. To do this, we adopt a Bayesian optimization protocol, iterating through the pair potential parameters against recently collected experimental data to probe a wider range of amino acid substitutions. We compare the pair potentials optimized for different proteins, as well as different physics-based loss functions to elucidate the factors influencing the development of transferable CG models. The insights gained in this work facilitate a deeper understanding of the applications of data-driven methods in understanding sequence-dependent phase separation of biomolecules.

Active learning to understand and predict the electronic behavior of peptides

Rajarshi Samajdar[1,2], Hassan Nadeem[3], Moeen Meigooni[2,4], Prateek Bansal[1], Nick Jackson[2,6], Martin Mosquera[7], Diwakar Shukla[1,2,3,7,9], Emad Tajkhorshid[2,3,5,6,7], Charles Schroeder[1,2,3,4,5,6]

[1]Department of Chemical and Biomolecular Engineering, University of Illinois at Urbana Champaign, Urbana, Illinois, 61801;
[2]Beckman Institute for Advanced Science and Technology, University of Illinois at Urbana Champaign, Urbana, Illinois, 61801;
[3]Department of Bioengineering, University of Illinois at Urbana Champaign, Urbana, Illinois, 61801;
[4]Center for Biophysics and Quantitative Biology, University of Illinois Urbana-Champaign, Urbana, IL 61801, United States;
[5]Department of Materials Science and Engineering, University of Illinois at Urbana Champaign, Urbana, Illinois, 61801;
[6]Department of Chemistry, University of Illinois at Urbana Champaign, Urbana, Illinois, 61801;
[7]Department of Chemistry and Biochemistry, Montana State University, Bozeman, MT 59717;
[8]Department of Biochemistry, University of Illinois at Urbana Champaign, Urbana, Illinois, 61801;
[9]Department of Plant Biology, University of Illinois at Urbana Champaign, Urbana, Illinois, 61801;

Abstract

Proteins play a key role in electron transport processes in biology, but the structure-function relationships governing the electronic properties of peptides are not fully understood. Our work aims to understand the connection between conformational flexibility, hierarchical structures, and electron transport behavior in peptides, which can be used to design and predict the electronic behavior of protein assemblies. In recent work, we studied the electronic properties of single peptides, and our results revealed a bimodal distribution molecular conductance across a relatively small set of amino acid sequences. MD simulations, Gaussian mixture modeling (GMM), and principal component analysis (PCA) were used to show that this two-state conductance behavior arises due to the conformational flexibility of peptide backbones. The high-conductance state arises due to a more defined secondary structure (beta turn or 3_{10} helices), and the low-conductance state is associated with the primary amino acid sequences. In this talk, I will discuss a large-scale collaborative approach that combines active learning with experiments, theory, and simulations to understand and predict the electronic properties of 324 tetrapeptides (M-X_1-X_2-M) and 5832 pentapeptides (M-X_1-X_2-X_3-M) over a large chemical sequence space. The N- and C-terminal residues are selected as methionine to facilitate binding to gold electrodes during single-molecule electronics experiments, and X_1, X_2 and X_3 represent all possible natural amino acids except methionine and cysteine. High-throughput MD simulations are performed using Folding@home with a cumulative sampling time of ~4 ms and ~17 ms for all tetra- and pentapeptides, respectively. Based on MD results, a subset of single-molecule experiments is performed on sequences that are predicted to show drastic differences in secondary structure formation. Experimental results are used in combination with MD to develop an active learning model utilizing classical and physics-inspired deep architectures that accurately predict molecular conductance for the entire tetra- and pentapeptide chemical space. Broadly, our work will demonstrate the ability to predict the electronic properties of folded molecular architectures based on the monomer sequence of amino acids. Overall, this work will provide new insights into understanding the role of higher order assembled structures on biological charge transport, in addition to showcasing the predictive power of a combined modeling and experimental approach to molecular electronics.

394e A Deep Learning-Driven Molecular Simulation Method to Explore the Folding Landscape of an RNA Stem-Loop

Ayush Gupta, Heng Ma, Arvind Ramanathan, Gül Zerze

The folding and unfolding of RNA stem-loops are critical biological processes; however, their computational studies are often hampered by the ruggedness of their folding landscape, necessitating long simulation times at the atomistic scale. Here, we adapted DeepDriveMD (DDMD), an advanced deep learning-driven sampling technique originally developed for protein folding, to address the challenges of RNA stem-loop folding. Although tempering- and order parameter-based techniques are commonly used for similar rare event problems, the computational costs and/or the need for a priori knowledge about the system often present a challenge in their effective use. DDMD overcomes these challenges by adaptively learning from an ensemble of running MD simulations using generic contact maps as the raw input. DeepDriveMD enables on-the-fly learning of a low-dimensional latent representation and guides the simulation toward the undersampled regions while optimizing the resources to explore the relevant parts of the phase space. We observed that DDMD estimates the free energy landscape of the RNA stem-loop reasonably well at room temperature. Our simulation framework runs at a constant temperature without external biasing potential, hence preserving the information of transition rates, with a computational cost much lower than that of the simulations performed with external biasing potentials. We also introduced a reweighting strategy for obtaining unbiased free energy surfaces and presented a qualitative analysis of the latent space. This analysis showed that the latent space captures the relevant slow degrees of freedom for the RNA folding problem of interest. We also identified the relevant stable and metastable states by using the learnt latent space. We will also outline how different parameters are selected and optimized to adapt DDMD for this system. We believe this presentation will help new users adapt this technique for the rare-event sampling problems of their interest.

394f Active Learning Cycle to Design Stable Collagen-Mimetic Peptoid Triple Helices

Alex Berlaga, Renyu Zheng, Diya Gandhi, Junhee Lee, Chunlong Chen, Andrew Ferguson

Peptoids (N-substituted glycines) are a novel, versatile class of synthetic sequence-defined polymers that are biomimetic, biocompatible, and easy to synthesize. Their secondary α-carbon and substituted nitrogen leads to flexible, tunable secondary structures and resistance to proteolysis. Incorporating peptoid residues in polymers has led to novel antimicrobial agents, toxic chemical detectors in cells, radioactive metal stabilizers, and many other applications in biological and materials science.

Collagen-Mimetic Peptides (CMPs) are short synthetic analogs to full-length natural collagen with applications in engineered tissues, biofilms, and sutures. Akin to natural collagen, CMPs are characterized by a triple-helix tertiary structure and often contain a large number of N-substituted proline residues. The apparent necessity of nitrogen substitution in CMPs introduces the possibility of hyperstable collagen-mimetic peptoid triple helices. Peptoid CMPs offer the potential to engineer novel biomaterials with attractive properties such as resistance to enzymatic degradation, optimized durability, and tunable bioactivities. The peptoid sequence-structure relationship is, compared to peptides, relatively poorly understood, making it challenging to prospectively identify which peptoid sequences are likely to assemble into stable and tunable peptoid CMPs.

In this work, we employed high-throughput simulation and active learning to computationally design and experimentally test hyperstable peptoid CMPs. Our computational design loop employs simulation-based assessment of the peptoid CMP melting temperature within an active learning cycle for machine learning-guided sequence design. Experimentally, the top candidates are synthesized, and their structure and stability is assessed using circular dichroism spectroscopy. Using this workflow, we demonstrate the design and production of the first highly thermostable peptoid CMPs with a range of future biomaterials applications.

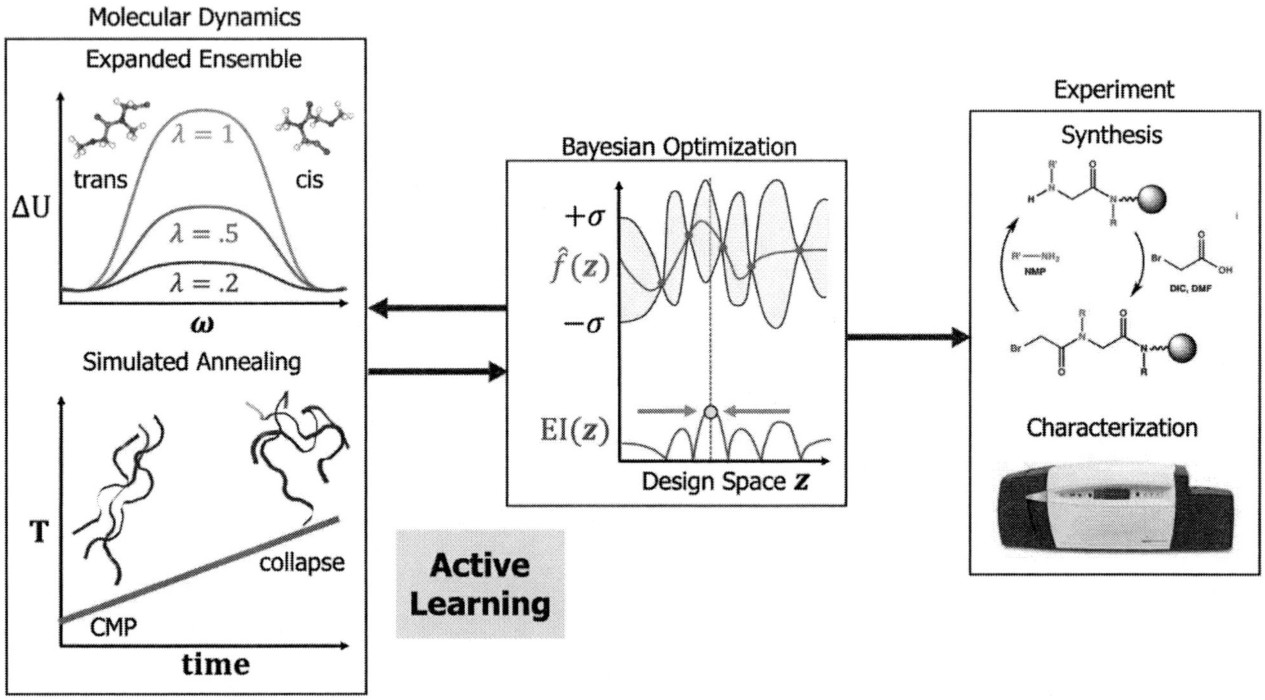

394g Deep Learning-Guided Anti-Bacterial Peptide Design

Camille Bilodeau, Daniel Garzon

Antibacterial peptides, a promising approach to treating a growing number of antibiotic-resistant diseases, often kill bacterial cells by binding to and altering the cell membrane. While antibacterial peptides typically leverage a combination of hydrophobic and electrostatic interactions to modulate peptide-cell membrane interactions, the role that peptide structure plays in governing potency is poorly understood. Neural networks offer a promising approach to learning this structure-function relationship. Unfortunately, publicly available datasets for these anti-bacterial activity are "data-poor" containing only 10^2-10^3 datapoints, limiting their utility in training neural networks.

In this work, we present a new transfer learning approach for anti-bacterial activity that involves first build a generalized, multi-task graph network to simultaneously predict peptide retention times on RPLC, HILIC, and IEX columns, tasks for which large (~10^5 datapoints) datasets are available publicly. We then finetune our model on a small (~10^3), publicly available antibacterial dataset. Finally, we apply our model to discover new antibacterial peptides by further refining our model on a small, internally generated, standardized dataset of point mutants. We demonstrate our model's ability to predict the outcome of new point mutations and apply our model to designing new candidate anti-bacterial peptides.

394h Active Learning for Polymer Blend Miscibility with Flory-Huggins Theory

Steven G. Arturo, Kaoru Aou, Jillian Emerson, Kathryn Grzesiak, Arjita Kulshreshtha, Paul M. Mwasame, Xiujiao Qiu, Clyde Fare, Jed W. Pitera, Edward Pyzer-Knapp

Polymer blends are increasingly being used for engineered solutions in materials development. Miscibility of polymer blends imparts an equal distribution of individual components throughout the volume of the blend, leading to a more uniform distribution in the material after formulation, processing, or cure. Given that three or more polymers can exist in such a blend, the search for compositions that are miscible demands numerous experiments. Active learning to guide experiments is done using Flory-Huggins theory as a surrogate model to map miscible and immiscible regions. Experiments are chosen using surrogate model uncertainty. The method is shown to greatly decrease the number of experiments needed to generate ternary phase diagrams for blends of homopolymers relative to space-filling methods.

394i Property-Guided Design of Topologically Complex Polymers Using Generative Modeling

Shengli Jiang, Adji Bousso Dieng, Michael Webb

The topology of a polymer chain can substantially influence their properties and those of derivative materials [1-2]. Establishing quantitative relationships between polymer topology and material properties remains challenging. Both experimental and computational investigations have enhanced the understanding of how polymer topology influences properties, yet such efforts are labor-intensive or computationally expensive, restricting studies to a small set of systems.

In recent years, efforts to design materials with target structural and functional properties have been increasingly enhanced by data-driven machine learning (ML) techniques [3-4]. However, successful applications of ML approaches for the design of polymers, or other soft materials, are just now emerging due to data scarcity, property complexity, and other factors. Nearly all demonstrations of ML over polymeric materials are restricted to simple linear architectures, yet the fact that polymers can possess complex architectures that dictate emergent properties is one of the motivating factors for their consideration as design constructs.

To address these limitations, we create a multi-task variational autoencoder (VAE) to generate polymers with specified topology and desired characteristics [4]. This model is developed with a dataset containing coarse-grained molecular dynamics (MD) data for 1,342 polymers. These polymers feature a variety of technologically relevant architectures such as star, comb, branch, linear, cyclic, and dendrimer structures, and cover a wide range of molecular weights. Input and encoding strategies are evaluated by training several models that aim to reconstruct the polymer topology and also perform auxiliary tasks of estimating the characteristic size (i.e., average squared radius of gyration) of the polymer and classifying its topology. Specifically, graph-derived topological descriptors (e.g., graph diameter, algebraic connectivity) and graph-explicit features are encoded and integrated to create a probabilistic latent space with physical interpretability. We leverage the generative modeling framework to produce sets of topologically diverse polymers that exhibit the same characteristic size in dilute solution. The rheological behaviors of these generated polymers, such as shear viscosity and viscoelastic moduli, are subsequently analyzed at finite concentrations. Overall, this study presents a scalable generative model for designing topologically diverse polymers with desired properties as an alternative to experiments or simulations.

[1] Khabaz, F. and Khare, R. Effect of chain architecture on the size, shape, and intrinsic viscosity of chains in polymer solutions: a molecular simulation study. The Journal of chemical physics 141 21, 214904 (2014).

[2] Wijesinghe, S., Perahia, D. and Grest, G. S. Polymer topology effects on dynamics of comb polymer melts. Macromolecules 51, 7621–7628 (2018).

[3] Gómez-Bombarelli, R. et al. Automatic chemical design using a data-driven continuous representation of molecules. ACS central science 4, 268–276 (2018).

[4] Kosuri, S. et al. Machine-assisted discovery of chondroitinase abc complexes toward sustained neural regeneration. Advanced Healthcare Materials 11, 2102101 (2022).

[5] Jiang, S., Dieng, A. B. and Webb, M. A. Property-guided generation of complex polymer topologies using variational autoencoders. ChemRxiv (2024).

394j Data Driven Machine Learning Based Inverse Polymer Membrane Design for CO_2 separation

Yasemin Basdogan, Zhen-Gang Wang

Designing polymer membranes with high gas permeability and selectivity presents a multifaceted challenge due to the inherent trade-off between these properties. This study introduces a data-driven, machine learning (ML)-based genetic algorithm aimed at the design of polymer membranes for CO_2 separation from N_2 and O_2. By utilizing literature data on the permeability of three gases, we generate Simplified Molecular-Input Line-Entry System (SMILES) representations of polymers based on their repeating units. Moreover, we leverage two fingerprinting techniques mentioned in commonly in the literature: one employs hash-based fingerprints, while the other uses substructure keys-based fingerprints. Our results indicate that hash-based fingerprints produce lower predictive errors in test sets compared to substructure keys-based fingerprints. Subsequently, we develop multiple ML models with hash-based fingerprints to predict the permeabilities and selectivities of gases in polymer membranes, demonstrating that regression models using tree-based ML algorithms are most effective in predicting gas permeability. We then integrate our 'forward ML model'—which predicts gas permeability and selectivity—with a generative algorithm for inverse design, allowing for a more effective exploration of the polymer material space. Genetic algorithms (GAs) are utilized to iteratively enhance solutions to the complex design problem, often navigating the vast and intricate search spaces more efficiently than traditional optimization methods. The initial gene pool is formed using the 'Breaking of Retrosynthetically Interesting Chemical Substructures' (BRICS) algorithm, yielding 79 unique genes from 780 reference polymers. To start the GA, we create 100 parent polymers with 4 genes in each monomer unit for the first generation. Fifteen families with the smallest Tanimoto similarity scores are then constructed, each with three parent polymers, and crossover operations are conducted to vary their chemical building block sequences, creating 12 offspring polymers per family. Each offspring's viability is assessed using a fitness function, which is pivotal in driving the evolutionary process of the GA. Running the GA for 100 generations resulted in more than 16,000 new polymer structures, among which promising candidates for CO_2/N_2 and CO_2/O_2 separations were identified. The top 100 polymers are predicted to possess high glass transition temperatures, with approximately 20% featuring pyridine functionalities. This data-driven ML approach to inverse polymer design is a potent tool that is applicable to designing polymers for any scenario requiring constrained optimization. Future work will concentrate on employing graph representation along with message passing graph neural networks to more accurately represent polymer complexity.

394k Predicting Vapor Liquid Equilibrium Curves of Acid Gas Absorption Using Machine Learning and Molecular Simulations

Frederick de Meyer, Kirill Klimenko, H. Mert Polat, Celine Houriez, Othonas A. Moultos, Alexander Varnek, Thijs J. H. Vlugt, Eric Cloarec

In the search for new solvents to remove acid gas (CO_2, H_2S, etc.) the use of molecular simulations and chemoinformatics is challenging because of the complexity of the absorption process, and, more specifically, due to its reactive nature. This talk will highlight several of our recent significant advances in computer-aided (mainly machine learning) developments in the field of absorption. The study case is absorption of CO2 and H2S in aqueous MDEA, a widely used industrial solvent.

The key thermodynamic property of acid gas removal is the acid gas-solvent Vapor-Liquid Equilibrium (VLE). These isotherms reflect the amount of acid gas absorbed for a given partial pressure of acid gas at equilibrium conditions. Measuring VLE curves is a time-intensive endeavor necessitating numerous repetitions for various temperatures and solvent compositions. Existing theoretical models for VLE prediction (e.g. e-NRTL) require the knowledge of a wide range of thermodynamic parameters of the studied pure, binary, ternary, etc. systems.

As a solution to this resource-intensive process, machine learning (ML) modeling has been employed to predict acid gas vapor pressure based on its concentration in the absorbent, temperature, and MDEA concentration in H2O. Data was collected from 15 articles, resulting in 512 data points (61 isotherms) for CO2 in MDEA+H2O and 239 datapoints (34 isotherms) for H2S in MDEA+H2O. Data analysis techniques were applied to the available data, and the resulting models were evaluated for their robustness and predictive capability. The model was used to predict new experimental data.The prediction error was deemed acceptable. A methodology is developed to eliminate suspicious experimental data, as well as unphysical trends. The accuracy of the model is also compared with a standard thermodynamic model fitted against experimental data and with the molecular simulation predictions presented in the next paragraph.

We will also briefly present the open-source chemical reaction equilibrium solver (CASpy, https://github.com/omoultosEthTuDelft/CASpy) developed to compute the concentration of species in any reactive liquid-phase absorption system. As a case study, we computed the CO_2 and H_2S absorption isotherm and speciation in an aqueous MDEA solution at 313.15 K and compared the results with available data from the literature. The results show that the computed CO_2 isotherms and speciations are in good agreement with experimental data, demonstrating the accuracy and the precision of our solver.

Selected publications:

1. Mert Polat, Frédérick de Meyer, Céline Houriez, Othonas A. Moultos, Thijs J.H. Vlugt, *Solving Chemical Absorption Equilibria Using Free Energy and Quantum Chemistry Calculations: Methodology, Limitations, and New Open-Source Software*, Journal of Chemical Theory and Computation, 19, 2616-2629, 2023.
2. Mert Polat, Hirad S. Salehi, Remco Hens, Dominika O. Wasik, Ahmadreza Rahbari, Frédérick de Meyer, Céline Houriez, Christophe Coquelet, Sofia Calero, David Dubbeldam, Othonas A. Moultos, Thijs J.H. Vlugt, *New features of the open source Monte Carlo software brick-CFCMC: Thermodynamic integration and hybrid trial moves*, Journal of Chemical Information and Modeling, 61 (8), 3752-3757, 2021.

394l A Data Science Framework for the Analysis of Ion Transport Mechanisms in Ionic Liquids

Elvis Umaña, Ryan Cashen, Matthew Gebbie, Victor M. Zavala

Ionic liquids are increasingly investigated for electrochemical energy storage as an alternative to flammable organic electrolytes due to their high stability and tunable chemical structures [1]. However, the strong intermolecular interactions which stabilize these molecules produce highly correlated ionic environments that complicate ionic liquid electrolyte design. The transport models commonly used to model ion transport in liquid electrolytes have led to a framework for maximizing conductivity by minimizing viscosity in ionic liquids [2]. Yet, recent studies show that many ionic liquids deviate from hydrodynamic behavior and instead suggest the presence of ion-hopping transport mechanisms in ionic liquids [3, 4].

To overcome the limitations of theory in understanding ionic liquid design, this work builds upon classical theory frameworks by using machine learning to identify structural and energetic motifs important for describing ion transport. We merge databases of experimentally measured properties and simulated molecular features for 218 ionic liquids. We then use these molecular descriptors in a machine learning model to correct hydrodynamic transport model conductivity predictions and gain insight to the physical origin of non-hydrodynamic ion transport behavior in ILs. Further, this work provides an in-depth analysis of the capacity for various ionic liquid information (structure, energetics, theory, and other properties) to predict ionic liquid ion transport. Intriguingly, we find that individual ion molecular features can predict some materials properties, such as conductivity, while failing to predict other properties that rely on larger length scale events, such as viscous dissipation.

Overall, this new framework provides a new avenue for identifying ionic liquid candidates with desirable properties while learning insights into the material property's physical origin.

1. Baskin, I., A. Epshtein, and Y. Ein-Eli, *Benchmarking machine learning methods for modeling physical properties of ionic liquids.* Journal of Molecular Liquids, 2022. 351.
2. Nordness, O. and J.F. Brennecke, *Ion Dissociation in Ionic Liquids and Ionic Liquid Solutions.* Chemical Reviews, 2020. 120(23): p. 12873-12902.
3. Cashen, R.K., et al., *Bridging Database and Experimental Analysis to Reveal Super-hydrodynamic Conductivity Scaling Regimes in Ionic Liquids.* J Phys Chem B, 2022. 126(32): p. 6039-6051.

575a Seeing the Invisible: In Nucleation (and in society)

Sapna Sarupria

Crystallization is consequential to many applications, including pharmaceutical production, flow assurance, and climate modeling. Effective control over crystallization relies on understanding the possible structures that can form during nucleation and growth. Molecular simulations allow a more fine-grained approach to discovering important, though possibly short-lived, intermediate structures, but their characterization from atomic coordinates is often difficult. We combine general features of the local atomic arrangements with a deep learning model to discover the unique structures that form during crystal nucleation. While many previous mechanistic studies have relied on features that describe the entire crystal nucleus, such as its size, shape, and composition, we focus on the evolution of the atoms involved in the formation of the nucleus in the feature space to describe nucleation processes. We find that this approach reveals structures not found by traditional approaches, and also enables us to classify nucleation pathways. Hence, we are able to shed light on previously "invisible" structures. This has implications in understanding and controlling the long-standing problem of polymorph selection during crystallization. In my talk, I will describe this method and briefly discuss invisible disabilities and what we can do as a community to continue being inclusive and fostering diverse talents and skills.

575b A Model for Chirality-Driven Phase Transitions

Pablo Debenedetti

A chiral tetramer model will be introduced. It is inspired by the geometry of the smallest and simplest chiral molecule, hydrogen peroxide. The model exhibits rich fluid-phase behavior, including chiral symmetry breaking. The underlying thermodynamics and phase transition kinetics will be discussed, as well as some possible implications for the origin of biological homochirality.

575c Multiscale Modeling of Hierarchical Self-Assembly Towards Advanced Functional Biomaterials

Mingfei Zhao

Hierarchical self-assembly provides a pathway for the bottom-up creation of nanomaterials, offering specific structures and functions that are challenging to achieve through top-down methods. Multiscale computational modeling emerges as a powerful tool to understand and control the intricate assembly processes across different length and time scales. This presentation highlights the effectiveness of computational modeling in understanding and directing self-assembly dynamics, with a focus on mesoscopic and microscopic systems.

Starting with the mesoscale examination of particle assembly within evaporating droplets, I integrated the lattice Boltzmann method for multiphase fluid dynamics with Brownian dynamics for particle behavior. These simulations provided profound insights into particle assembly, particularly at the liquid-vapor interface, enhancing the capabilities for sophisticated fabrications. Moving to smaller scales, my research explored the hierarchical self-assembly of peptoids using molecular dynamics simulations. Peptoids, known for their exceptional biocompatibility and stability, offer promising applications in antimicrobials, drugs, and catalysts. Pioneering simulation efforts revealed critical insights into the structures and properties of peptoid-based nanomaterials. Additionally, multiscale simulations were also developed to investigate interactions between the HIV-1 fusion peptide and complex cell membranes. This work provides critical biophysical understanding to inspire drug discovery effort to inhibit HIV-1 cell entry.

This presentation underscores the significance of multiscale simulations in building foundational understanding and envisions future directions for the rational design of advanced functional biomaterials.

575d Leveraging Quantum Mechanics for Soft Materials Design

Nicholas Jackson

The dramatic progress of soft materials to date has been driven by the use of models based in classical physics. While the successes of these models are beyond reproach, some of the most pressing societal challenges (e.g. sustainability, flexible electronics, energy storage) necessitate the incorporation of quantum mechanics (QM) into the design of soft materials. However, the need for QM predictions applied at the length and time scales characteristic of soft materials represents a substantial computational challenge for which new approaches are needed. I will describe recent work extending and applying QM prediction methods to soft materials design spanning multiple materials classes spanning bond rupture in thermosets, charge conductivity in semiconducting polymers and liquid crystals, and the characterization of the fundamental physical properties across a broad macromolecular genome. Integral to these cases will be the application of advanced machine learning methodologies with new strategies in both "top-down" and "bottom-up" coarse-grained modeling to incorporate electronic predictions.

575e Developing Strategies for Polymer Redesign and Recycling Using Reaction Pathway Analysis

Linda J. Broadbelt

The current lack of sustainability of and the limited portfolio of recycling processes for synthetic polymers have posed serious threats to the environment. Approximately 90% of plastics are produced via fossil fuels, and over 150 million tonnes of plastics have been discarded in the ocean. Annual production of plastics is expected to reach over 1 billion tons in 2050, but the current manufacturing, consumption, and disposal schemes of fossil-based polymers follow an unsustainable framework. Using reaction pathway analysis, we are pursuing a portfolio of strategies for redesign and recycling of polymers for sustainability.

Pyrolysis is a promising method for resource recovery from plastic waste that is compatible with current petrochemical infrastructure that thermally converts polymers in the absence of oxygen into valuable chemical feedstocks and monomer. To provide further insight into polymer pyrolysis, a greater understanding of the mechanistic and kinetic details of the underlying reaction network is needed. To handle the complexity of mechanistic modeling of polymer degradation, we have developed both continuum and kinetic Monte Carlo (kMC) models. We have applied these approaches to study the degradation of polystyrene, polypropylene, polyisoprene, polystyrene peroxide and binary mixtures, and we are able to capture diverse experimental measures, including yields of individual low molecular weight products, as a function of reaction conditions. In order to solve the large models that are created, values of the rate coefficients for $O(10^5)$ reactions are typically required. The approach that we have developed to specify rate coefficients is hierarchical, based on a combination of literature values, estimation methods, and computational chemistry.

Alternatively, redesign efforts focusing on polymers that can be reused and recycled to monomers can lead to sustainable solutions for the plastics waste problem. One pathway to success is to identify bioprivileged molecules, biology-derived chemical intermediates that can be efficiently converted to a diversity of chemical products, including both novel molecules and drop-in replacements, and molecules emanating from them that can be used as monomers leading to recyclable polymers. We have developed a framework for molecule discovery and reaction pathway design that is automated and flexible and can be used to screen for bioprivileged candidates and target molecules. The application to discovery of known and novel monomers for poly(hydroxyurethanes) that are derived from biobased molecules and lead to recyclable materials will be discussed, and computational methods to evaluate the recyclability of different polymers will be outlined.

709g Discovery of New Surfactants Used in Firefighting Foams By Active Learning

Xin Wang, Hongwei Zhang, Seokgyun Ham, Rui Qiao

Aqueous Film Forming Foams (AFFF) used to suppress liquid fuel fires contain fluorinated surfactants that are being banned internationally due to their persistence and toxicity. New surfactant solutions for replacing AFFF are being sought, and fuel transport at the water-fuel interface has been identified as a potential key mechanism affecting fire extinguishment performance. Traditional design approaches, however, are limited in exploring the vast chemical space of surfactants. In this work, we initially collected a dataset containing 201 samples quantifying octane transport resistance of surfactants accumulated water-fuel interface generated by coarse-grained molecular dynamics simulations. The dataset encompasses a variety of surfactant types, including cationic, anionic, zwitterionic, non-ionic, sugar-based, and Gemini surfactants. For design purposes, we focused on hydrocarbon surfactants with identical tails, varying only the headgroups, as our previous work indicated that the surfactant tail has negligible impact on fuel transport. Subsequently, we employed an active learning protocol that combines a Graph Convolutional Network (GCN) model with Bayesian optimization to predict interfacial resistance and identify top-performing candidates efficiently. Notably, surfactants featuring beta-D-glucuronic acid heads show excellent performance in blocking octane transport; the resistance of their monolayer at interfaces is equivalent to that of ~ 3800 nm-thick water layer. In summary, this study not only enhances our understanding of fuel transport across the water-fuel interface but also identifies promising new surfactant solutions for fire suppression. Furthermore, the generalizable design framework we established can be extended to other surfactants, such as siloxane surfactants.

169a Exploring Effective Design Strategies in Fine-Tuning the Electronic, Transport, and Photophysical Properties in Cofs

Yuanhui Pan, Francesco Paesani

Covalent organic frameworks (COFs) have emerged as promising materials due to their highly tunable composition, structure, and physical properties, alongside their high porosity and large surface area. Despite their potential in gas separation, energy storage, catalysis, and optoelectronics, a unified understanding of the intricate relationship between composition, structure, intermolecular interactions, and electronic properties in COFs from an *ab initio* perspective remains elusive. Here, we present a systematic study that integrates multiple theoretical approaches, ranging from all-electron quantum chemistry to coarse-grained model Hamiltonians, specifically within the Density Functional Theory (DFT) and multi-particle Holstein formalism (MHF). Concretely, we find that functionalizing the linker with small functional groups in highly-symmetric lattices minimally alters the electronic structures, particularly when the backbone is dominant. Conversely, breaking symmetry in building blocks via functionalization, and if in turn leads to changes in lattice symmetry, proves to be an effective strategy for modulating electronic, transport, and photophysical properties. Our results provide physical insights as well as guiding principles for application-driven experimental discovery of novel COFs with enhanced photo-catalytic prowess. Finally, our methodology is an effective tool that enables disentangling the structure-property relationships in reticular materials.

169b Data Set and Data-Driven Models for Predicting Metal-Organic Framework Stability in Water and Harsh Environments

Gianmarco Terrones, Shih-Peng Huang, Matthew Rivera, Shuwen Yue, Alondra Hernandez, Heather Kulik

Metal-organic frameworks (MOFs) are porous materials with applications in gas separations and catalysis, but a lack of water stability often limits their practical use given the ubiquity of water in air and the environment. Consequently, it is necessary to predict whether a MOF is water-stable before investing time and resources into synthesis. While heuristics for designing water-stable MOFs such as hard-soft acid-base theory are well-known, these can lack generality and artificially limit the diversity of explored chemistry due to narrowly defined criteria. Thus, we instead use machine learning (ML) models trained on experimental MOF stability data to improve generality and utilize examples spanning MOF chemical space. In an improvement on previous ML efforts, we enlarge the available training data for MOF water stability prediction by over 400%, adding 911 MOFs with water stability labels assigned through manuscript analysis. The additional data is shown to improve ML model performance (test ROC-AUC > 0.8) over diverse chemistry for the prediction of both water stability and stability in harsher acidic conditions. We illustrate how the expanded data set and models can be used with previously developed activation stability models to run genetic algorithms that quickly screen ~10,000 MOFs to identify candidates with multivariate stability (i.e., for activation, in water, and in acid). Model analysis and genetic algorithm results uncover metal and geometry-specific design rules for synthesizing robust MOFs. The data set and ML models developed in this work, which we disseminate through an easy-to-use web interface, are expected to contribute toward the accelerated discovery of novel, water-stable MOFs for applications requiring water stability such as direct air gas capture and water treatment.

169c Prediction of pK_a in Different Solvents Via Deep Learning

Jonathan Zheng, Ivari Kaljurand, Sofja Tshepelevitsh, Ivo Leito, Thomas Nevolianis, Simon Müller, William Green

The acid dissociation constant, or pK_a, is important in many applications including drug discovery, chemical synthesis, and environmental studies. Most publicly available models and data compilations are developed for aqueous pKa predictions; resources for nonaqueous solvents are comparatively much rarer. Yet, pK_a values in different solvents can differ by orders of magnitudes.

The difference in pK_a of a compound between two solvents can be calculated using solvation models. These calculations can be used with aqueous pK_a data, as well as an "anchor" acidity value in the desired solvent, to compute pK_a in nonaqueous solvents. Previously, we have shown that this method can be used with the COSMO-RS solvation model to compute dissociation constants with mean absolute errors (MAEs) less than 1 log unit in several solvents, with good performance for solutes including small molecules such as amino acids, and neurotransmitter derivative molecules.

In this work, we leverage this method to develop a dataset of computed pKa values. We introduce previously-unpublished experimental data, including datasets that have recently been, or are in the process of being, critically reviewed by IUPAC. We combine these synthetic and experimental datasets to develop a large corpus of training data, which is used to train a directed message-passing neural network (D-MPNN) model. Finally, we evaluate the model performance, comparing against the performance of other nonaqueous pKa models.

169d Exploring the Role of Functional Groups and Nanoconfinement on the Structural and Dynamical Properties of Water and Ions inside Metal-Organic Frameworks

Akash Ball, Shuwen Yue, Heather Kulik

Metal-organic frameworks (MOFs) are potential candidates for next-generation membranes with high ion selectivity due to the presence of specific ion binding sites created by incorporating functional groups into the organic linkers of MOFs. To design MOF membranes with high water flux and excellent ion-selectivity performance, a fundamental understanding of the structural and dynamical properties of water and ions inside MOFs is essential. Although a handful of previous studies have explored a few such structural and dynamical properties of water inside MOFs, no studies have been carried out to identify how functional groups affect these properties of water and ions. Here, we take a computational approach to determine the effect of different functional groups and confinement on the water/ion coordination environment and the transport inside Zr-based UiO-66 MOF. First, we perform classical molecular dynamics (MD) simulations of water and ions inside functionalized UiO-66 MOFs, which reveal that water coordination structure and ion hydration structures are significantly different under nanoconfinement than bulk water. The MD simulations also indicate that water undergoes both Fickian diffusion and sub-diffusion inside UiO-66, and the mode of diffusion depends on the functional group size and the number of functional groups on each linker. Next, we performed density functional theory (DFT) calculations to determine the binding affinity of water and ions with different functional groups. The DFT calculations show that water and ion diffusion coefficients strongly correlate with binding affinity. Finally, we extend our analysis to other MOFs to validate the generalizability of our results and provide design guidelines for MOF-based membranes with fast water transport and unprecedented ion-selectivity for applications in battery recycling, wastewater treatment, and rare earth element recovery.

169e Integrating Off-Lattice Kinetic Monte Carlo with Molecular Dynamics for Modeling Polyvinyl Chloride Dehydrochlorination

Feranmi Olowookere, Heath Turner

Many plastic wastes, including polyvinyl chloride (PVC), contain considerable repositories of molecular value. These resources can potentially be accessed via depolymerization mechanisms; we aspire to perform these transformations using scalable techniques at mild conditions. In this study, a three-dimensional off-lattice kinetic Monte Carlo-molecular dynamics (KMC-MD) framework [*Comp. Mat. Sci.* 229, 112421 (2023)] is developed to investigate the dehydrochlorination/conjugation transformation of PVC in sodium hydroxide (NaOH) with atomistic resolutions at experimental timescales ($10^3 – 10^6$ s). Our framework enables a comprehensive examination of the competing reaction pathways and molecular-scale changes influenced by various solvents (acetone, ethylene glycol, triethylene glycol, tetrahydrofuran, and bio-derived solvents), as well as the influence of varying molecular weight distributions, concentrations, and temperatures. The algorithm simulates bond cleavage and formation during the KMC stages, whereas the MD stage is dedicated to the relaxation and thermalization of the PVC-NaOH-solvent system. The framework allows us to capture important configurational aspects (mixing, correlations, clustering, etc.) that are not accessible with a traditional microkinetic model. Our KMC-MD approach provides an exceptionally powerful tool for elucidating the complex reaction pathways, while simultaneously gaining insights into the molecular changes inherent to the depolymerization processes. Our technique allows us to perform direct experimental benchmarking at experimental timescales, paving the way for more effective upcycling of polymer wastes into valuable chemicals under mild conditions.

169f Utilizing an Integrated Experimental and *in silico* Approach to Engineer Cross Reactive Antibodies Against VEGFA and PlGF2 in Pediatric Glioblastoma

Emma C. Aldrich, Samuel Blackman, Jennifer Maynard, Kayla G. Sprenger

Pediatric glioblastoma is characterized by the overexpression of growth factors that target angiogenic pathways in tumors, presenting a unique opportunity for therapeutic design. In healthy cells, Vascular Endothelial Growth Factor-A (VEGFA) contributes to vascular permeability and angiogenesis, but overexpression in cancers is known to contribute to tumor angiogenesis. Meanwhile, Placental Growth Factor 2 (PlGF2) plays a similar role in placental angiogenesis and vascular development, but is also upregulated across many cancer types. Though progress has been made to design antibodies to block VEGFA interactions with its family of Vascular Endothelial Growth Factor Receptors (VEGFRs) and its co-receptor, Neuropilin 1 (NRP1), the introduction of anti-VEGFA antibodies can lead to therapeutic escape by expressing related PlGF2. Critically, sequence identity between growth factors can be leveraged to design cross-reactive antibodies to simultaneously target VEGFA and PlGF2 and their interactions with NRP1 and VEGFR1, presenting opportunities to improve specificity and versatility *in vivo* and expand clinical translatability. To this end, our experimental collaborators recently determined a suite of novel candidate antibodies and measured their affinity and specificity/cross-reactivity for VEGFA, PlGF2, and antibody-growth factor complexes with NRP1, VEGFR1, and heparin sulfates. This poster will describe our use of computational approaches, including static docking and molecular dynamics simulations, to elucidate the molecular-level interactions between the relevant growth factors and newly-derived antibodies. Additionally, it will highlight our efforts to identify the molecular basis for the apparent cross-reactivity of these antibodies, via assessing the effects of single point mutations introduced into the antibody sequences *in silico*. Finally, we describe our efforts to quantify competition between the growth factors and antibodies in the presence and absence of NRP1 and VEGFR1. Overall, characterization of this antibody suite informs angiogenic interactions between VEGFA and PlGF2 signaling in pediatric glioblastoma, leveraging *in silico* approaches to advance efforts toward robust and translatable cancer therapeutics.

169g Evaluating Suitability of the Chimes Machine-Learned Interatomic Model for Zeolite Materials

Sayed Ahmad Almohri, Rebecca Lindsey

Catalytic membrane reactors employing zeolite membranes hold promise for efficient ammonia production. Zeolite membranes act as selective barriers within the reactor, allowing for *in situ* separation of ammonia from the reaction mixture. This eliminates the need for a separate separation stage, improving process efficiency. [1], [2] Designing optimal zeolite membranes requires accurate simulations of complex chemical systems. However, traditional computational methods, such as density functional theory (DFT), can be prohibitively expensive for large systems while classical forcefield-based approaches may be inadequately accurate in the case of reacting systems. To bridge this capability gap, we develop a Chebyshev Interaction Model for Efficient Simulations (ChIMES) for simulating zeolite materials relevant to ammonia synthesis. ChIMES is a unique physics informed many-bodied machine learned interatomic potential and semi-automated fitting framework that combines the accuracy of DFT with the computational efficiency of molecular mechanics. ChIMES has been successfully applied to study the behavior of multiple chemical systems.[3], [4], [5] In this work, we explore the application of ChIMES to characterization and optimization of zeolite nanosheet membranes ChIMES by simulating zeolite systems for the design of our membranes, and discuss strategies for efficient model parametrization.

[1] H. Zhang *et al.*, "Open-Pore Two-Dimensional MFI Zeolite Nanosheets for the Fabrication of Hydrocarbon-Isomer-Selective Membranes on Porous Polymer Supports," *Angew Chem Int Ed Engl*, vol. 55, no. 25, pp. 7184–7187, Jun. 2016, doi: 10.1002/anie.201601135.

[2] P. Kumar *et al.*, "One-dimensional intergrowths in two-dimensional zeolite nanosheets and their effect on ultra-selective transport," *Nat. Mater.*, vol. 19, no. 4, Art. no. 4, Apr. 2020, doi: 10.1038/s41563-019-0581-3.

[3] R. K. Lindsey, L. E. Fried, and N. Goldman, "ChIMES: A Force Matched Potential with Explicit Three-Body Interactions for Molten Carbon," *J. Chem. Theory Comput.*, vol. 13, no. 12, pp. 6222–6229, Dec. 2017, doi: 10.1021/acs.jctc.7b00867.

[4] R. K. Lindsey, L. E. Fried, and N. Goldman, "Application of the ChIMES Force Field to Nonreactive Molecular Systems: Water at Ambient Conditions," *J. Chem. Theory Comput.*, vol. 15, no. 1, pp. 436–447, Jan. 2019, doi: 10.1021/acs.jctc.8b00831.

[5] R. K. Lindsey, N. Goldman, L. E. Fried, and S. Bastea, "Many-body reactive force field development for carbon condensation in C/O systems under extreme conditions," *The Journal of Chemical Physics*, vol. 153, no. 5, p. 054103, Aug. 2020, doi: 10.1063/5.0012840.

169h Computational and Experimental Studies of Hydrophoic, Nonaqueous, Nonvolatile, Low Viscous, and High CO_2 Absorption Chemical Solvents

Wei Shi, Robert Thompson, Jeffrey Culp, Surya Tiwari, Jan Steckel, Nicholas Siefert, David Hopkinson

Solvents with high CO_2 loadings in the intermediate CO_2 pressure range of 2-5 bar have lots of industrial applications, such as CO_2 capture from the steam methane reforming process. Recently, experimentalists at NETL have synthesized a new class of solvent, which contains NH and COO or NCO groups. This class of solvent was found to give high CO_2 loadings at the pressure range of 2-5 bar due to a reversible chemical reaction at the amine site with CO_2. In this presentation, we present the computational and experimental studies of this class of solvent.

Due to the complicated molecular structure for the EMA-NH-EMA solvent molecule, a python script code was developed to automatically generate 30-40 different initial structures, which were optimized by using the UFF classical force field, for both the reactants and products by using the RDKit package. These optimized initial configurations were then further optimized at the B3LYP/6-311++g(d,p) level of theory in the gas phase by using the Gaussian 09 software package. Simple scripts were developed to automate to conduct the quantum mechanical (QM) calculations and analyze the QM results obtained from Gaussian calculations. In addition, classical force field was developed for the solvent to predict its physical properties, such as vapor pressure obtained from the solvation free energy calculation by using the BAR method implemented in Gromacs, solvent viscosity, and CO_2 diffusivity, CO_2 and H_2 physical solubilities. QM calculations show that for the reactant EMA-NH-EMA, a configuration with intra-hydrogen bonding between H (NH) and O of (CO) exhibits the lowest energy, while an extended configuration without intra-hydrogen bonding network exhibits about 9 kJ/mol higher energy. For CO_2 reaction with EMA-NH-EMA to form carbamic acid (NCOOH), two typical configurations are possible. The first configuration is that H atom of NCOOH forms intra-hydrogen bond interaction with O of CO, and the second one is the extended configuration with NCOOH pointing outwards without intra-hydrogen bonding. In contrast to the reactant, for the reaction product, the second extended configuration is more stable and exhibits about 10 kJ/mol less energy than the first configuration. Molecular dynamics (MD) simulations using the second extended configuration for the product shows a high viscosity of 62.3 cP compared with the low viscosity of 7.8 cP for the neat solvent. The high viscosity of 62.3 cP was due to the inter-hydrogen bonding, which leads to CO_2 diffusivity to decrease by 3.4 times in the product compared with the neat reactant. In contrast, the experimental data does not suggest a significant viscosity increase upon CO_2 reaction with the EMA-NH-EMA solvent. This inconsistency for the product viscosity between simulations and experimental data suggests that the first intra-hydrogen bond configuration for the product will be present and simulations will be presented. Computations also suggest that the EMA-NH-EMA solvent exhibits a very low vapor pressure of 0.056 (8) Pa at 298 K, which is comparable to the nonvolatile Selexol solvent. Finally, CO_2 physical solubilities in both the neat solvent and the product have been calculated and they were found to be reasonably high (1.21-1.41 mol/L.MPa at 298 K), which is consistent with the experimental observation that CO_2 solubility still appreciably increases at elevated pressures instead of approaching a plateau region for a typical chemical solvent. H_2 physical solubilities in both the neat solvent and the product were also calculated and they were found to be favorably low (0.0199 - 0.02484 mol/L.MPa at 298 K) and they will be compared with experimental data. Temperature effects on CO_2 and H_2 physical solubilities will also be shown. Finally, we note that due to significant amounts of CH_2 groups in the solvent, this class of solvent is very hydrophobic.

169i Studying Depolymerization of Polyurethanes Using Reaction-Aware Deep-Learning Potentials

Lakshmi Yasodhara Ananthabhotla, Siddarth Achar, Karl Johnson

Polyurethanes are one of the most widely employed classes of polymers. Polymerization of polyurethanes (PUs) is a simple process facilitated by the formation of strong urethane bonds between the constituent moieties. This strong urethane linkage contributes to wide-ranging applications across a diverse spectrum, from hard insulating materials to soft foams. However, the high stability of the urethane bond makes it non-degradable in nature, making its disposal an environmental concern. This makes chemical depolymerization an important recovery method to mitigate the waste accumulation and reduce the environmental impact of this polymer. Our work focuses on developing a framework for efficiently training reaction-aware deep-learning potentials (DPs) that can be used to study many possible reactions and reaction pathways for PU depolymerization. Our aim is to implement the "reactive active-learning' formalism that will efficiently train reaction-aware machine learning potentials to study PU depolymerization. This implementation will also help us identify the reaction conditions that will improve the efficiency and selectivity of the process. Our initial work involved training bootstrap DPs using density functional theory data. Next steps include sampling possible bond breaking and bond making events to form products using the DPs and adding them back into the training data set to retrain the DPs.

169k Computational Modeling and Design of Self-Stratifying Colloidal Materials

Mayukh Kundu

Mixtures of colloidal particles suspended in a solvent can spontaneously form layered structures during fast solvent drying. This process, called self-stratification, can be leveraged to fabricate multilayered coatings (or supraparticles) from films (or droplets) in a single processing step. Such materials have applications in pressure-sensitive adhesives, abrasion-resistant coatings, drug encapsulation, and catalysis. The occurrence and extent of self-stratification are experimentally known to be sensitive to numerous physicochemical properties of the colloidal suspension and the processing conditions, but the process is poorly understood theoretically, and so is difficult to engineer. I am using computational modeling to understand the physics of self-stratification and then to design self-stratifying materials with targeted composition gradients.

Existing models for simulating self-stratification are computationally expensive or inaccurate. I have developed a better model for simulating the phenomena using dynamic density functional theory (DDFT). DDFT is a continuum model that is systematically formulated from particle-level interactions and dynamics. As such, it incorporates physics that would be present in particle-based simulations but can access much larger length scales and longer time scales. DDFT has two key inputs: a thermodynamic model (free-energy functional) and a dynamics model (mobility tensor). Different approximations of these inputs can be made, and the model can be made faster using the simplest models that give the desired accuracy. I systematically investigated approximations of both inputs to develop an accurate, efficient DDFT model for drying suspensions. I then coupled this model to an optimization strategy based on surrogate modeling to "inverse design" self-stratified coatings with targeted thickness and particle distribution. My work has the potential to reduce the time and resources required to create these novel materials in the laboratory.

169l The Tradeoff between Chemical Accuracy and Computational Cost: An Assessment of Thermochemical Prediction with Density Functional Theory

Rebecca Walters, Joseph S. Gomes

Density functional theory (DFT) methods are widely used for computing thermochemical properties such as non-covalent interactions, barrier heights, and reaction energies. The choice of a particular model chemistry, which includes selecting an appropriate exchange-correlation (XC) functional and basis set, is important for achieving chemical accuracy. There exist several classes of XC functionals, each experiencing some degree of tradeoff between chemical accuracy and computational complexity. Additionally, increasing the size of the basis set generally improves the accuracy of computed molecular properties at an additional computational cost. While there have been numerous benchmarking studies for density functionals, the majority only focus on chemical accuracy. We argue that it is necessary to also consider the compute wall time when selecting an appropriate model chemistry suitable for e.g. high-throughput computational screening or big data assembly.

In this work, we benchmark XC functional and basis set pairings using datasets covering a wide range of thermochemical prediction tasks, including non-covalent interactions (DES15K), barrier heights (BH9), reaction energies (BSE49), and an all-around thermochemistry benchmark (GMTKN55). We investigate the use of some empirical correction methods which address known deficiencies in DFT, such as incomplete basis set error, electron self-interaction error, and the underestimation of Van der Waals interactions. We show that smaller basis sets paired with the DFT-C correction are capable of achieving near chemical accuracy on a surprising number of property prediction calculations. We find that while XC functionals with high computational complexity consistently perform better, XC functionals with relatively lower complexity can perform similarly when paired with the appropriate empirical corrections. We note trends of chemical accuracy of density functionals and give several recommendations on model chemistry choice with respect to necessary compute time and type of chemistry.

169m Using ML to Determine the Optimal Set of Operational Parameters of the RO System in a Regional Water Treatment Plant

Arash Tayyebi, Ali Alshami

Operating RO systems presents challenges that involve handling a large amount of data, monitoring various variables, and conducting statistical analysis before sending water for treatment. Operators must monitor feed water characteristics such as conductivity, oxidation reduction potential (ORP), total suspended solids (TSS), turbidity, and chemical oxygen demand (COD), as well as the operational parameters of RO membrane feed such as feed pressure, flow rate, and temperature, as the most impactful factors affecting RO membrane performance to optimize the operational parameters like salt passage, permeate flow rate, and pressure difference across membranes. Data from individual RO membrane units must be collected and normalized, providing criteria for choosing the best treatment methods and minimizing costs.

Currently, many facilities still rely on manual SCADA data collection and infrequent review of maintenance logs, a process that can take up to weeks. Furthermore, water varies in hybrid plants such as the one in this study, making it more difficult to select the best treatment method for an effective RO facility's operation and maintenance strategy. ML models can be employed to better monitor data and predict the optimal set of operational parameters, maximizing membrane performance in an RO system and enabling rapid response and mitigation and safety efforts.

This study aims to develop a predictive model capable of assessing the impact of feed characteristics (including conductivity, ORP, TSS, turbidity, and COD) and operational parameters of RO membrane feed (such as feed pressure, flow rate, and temperature) on the performance of RO membranes, focusing on parameters like salt passage, permeate flow rate, and pressure difference across membranes. The performances of different machine learning methods will be compared, and the model with the highest accuracy will be used to predict the best treatment method by adjusting water quality and determining the most influential factors on RO membrane performance. Subsequently, an optimization model will be developed to determine the optimal set of operational parameters for the RO system, aiming to maximize membrane performance.

169n Surface-Centered Approach for Characterization and Prediction of Protein-Membrane Interactions

ByungUk Park, Reid Van Lehn

Peripheral membrane proteins (PMPs) are the membrane proteins that transiently bind to the membrane via surface interactions and participate in crucial cellular processes such as signaling cascades and lipid metabolism. Characterizing and predicting the membrane-binding interfaces of PMPs is important for developing novel therapies that disrupt PMP-membrane interactions or for rationally designing protein scaffolds to mediate interactions at target membranes. Predicting the membrane-binding interface is a challenging task, as the physicochemical properties of the interface are similar to those of nonbinding interfaces. The available structures of peripheral proteins are also obtained from their soluble states since stabilizing the membrane-bound state is difficult with contemporary experimental techniques. Thus, most mechanistic and energetic studies of PMPs have been performed using molecular dynamics simulations, but the time required to sample the whole protein-membrane interaction (PMI) process is tremendously large in case of all atomistic-level simulations. Therefore, developing methods to achieve fast predictions of PMP membrane-binding interfaces will broaden our understanding of the PMI process and can be employed for various purposes including the rational design of membrane-targeting drugs, biosensors with enhanced binding specificity and sensitivity, and antifouling coating for medical or marine industry.

This study seeks to understand and characterize the general surface properties of the interfaces between protein and membrane, then leverage this surface-centered understanding to predict PMP membrane-binding interfaces. To address this goal, we define numerical arrays, or fingerprints, that capture geometric and chemical features of the protein and membrane surfaces. These vectors are then used as input for a geometric deep learning model to predict the interaction between protein and membrane of interest and identify the specific regions of the PMP surface that correspond to the membrane-binding interface. To increase the accuracy of the approach and capture the flexibility associated with some membrane-binding domains, we perform molecular dynamics simulations to sample conformational dynamics as additional model input. We show the capability of this model to discriminate binding interfaces for a dataset consisting of over 1000 PMPs. This approach enables membrane-binding interface predictions and will further enable new mechanistic interrogation of binding modes to produce fundamental physical insight into the mechanisms that PMPs use to interact with membranes.

169o Multimodal Language and Graph Learning of Adsorption Energy Prediction

Janghoon Ock, Rishikesh Magar, Akshay Antony, Amir Barati Farimani

In machine learning models designed for catalyst screening, adsorption energy is often the primary modeling target because it acts as a key descriptor of reactivity. This energy is determined by identifying the minimum energy among various adsorption configurations on the catalytic surface. This process necessitates the models to differentiate subtle differences in energy levels and structures across the adsorption configurations. Graph Neural Networks (GNNs) have emerged as the favored method for modeling atomic systems. However, the graph representation requires exact atomic spatial coordinates and often lacks human interpretability. Conversely, recent advances in language models have extended their utility in predicting catalytic properties, presenting an alternative to complex graph representations. These models are proficient in processing human-readable text data, facilitating the seamless integration of observable features. However, their energy prediction accuracy, with a mean absolute error (MAE) of 0.71 eV, is insufficient for discerning the subtle differences in energy that could vary from 0.1 to 0.3 eV between adsorption configurations. This study aims to address this limitation in accuracy by introducing a self-supervised multimodal learning strategy, termed graph-assisted pretraining. This technique employs a contrastive loss between embeddings from graph and text encoders, thereby enabling the knowledge transfer from graph to text embeddings by aligning them across the same systems. This method has notably enhanced the MAE to 0.35 eV, achieving accuracy on par with DimeNet++ while utilizing merely 0.4% of the training data. Moreover, the Transformer encoder of the language model enables an examination of feature focus through its attention scores. This indicates that our multimodal training approach effectively redirects the model's focus toward relevant adsorption configurations and adsorbate-related features. This study lays the groundwork for merging different modality representations in adsorbate-catalyst system studies.

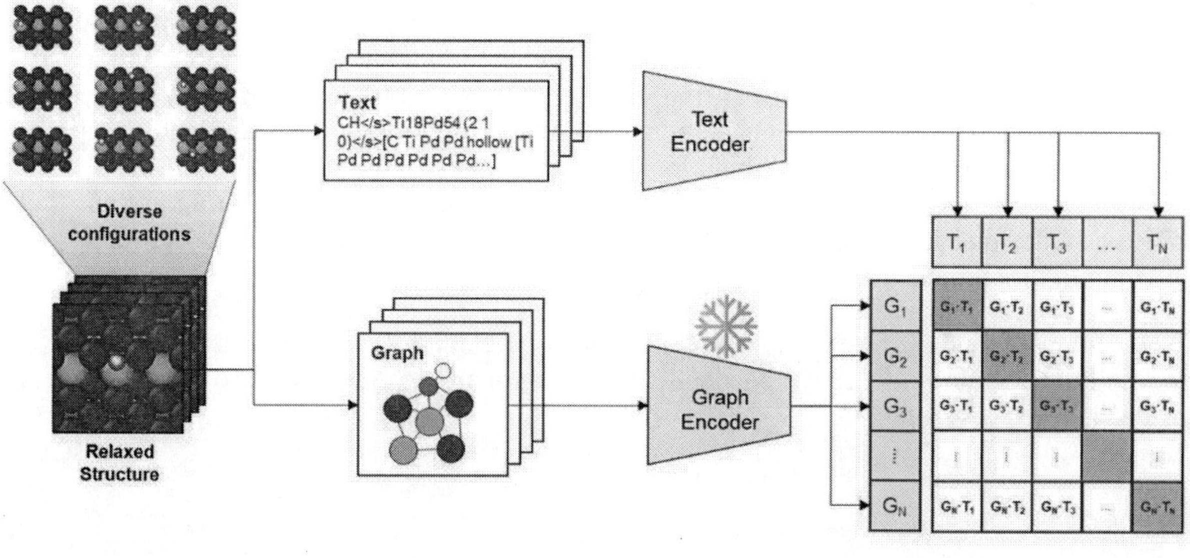

169p Ligand Lipophilicity and Architecture Influence Mechanisms and Thermodynamics of Nanoparticle Adsorption to Lipid Bilayers

Carlos Huang-Zhu, Reid Van Lehn

The design of ligand-coated nanoparticles that can translocate across biological membranes with minimal membrane disruption has been challenging due to the lack of knowledge on cell-penetrating mechanisms. Ligand lipophilicity has been studied as the primary descriptor for predicting biological interactions for this class of nanoparticles, particularly in drug discovery. However, the molecular mechanisms and thermodynamic driving forces through which these ligands interact with biological membranes are yet to be fully explained through experiments. Furthermore, the timescales of experimental synthesis and characterization have posed challenges by limiting the design space for these nanomaterials as biomedical agents.

In this work[1], we employ coarse-grained molecular dynamics simulations to study the adsorption of ligand-coated gold nanoparticles to single-component bilayers. The ligands studied possess a cationic linear alkyl end group that varies in lipophilicity through alkyl chain length. We used path variables and the string method with swarms-of-trajectories to define the minimum free energy path for adsorption as a function of two collective variables. We define the distance along this path as a reaction coordinate and compute a corresponding potential of mean force using umbrella sampling for each nanoparticle. We show that adsorption to a simple bilayer is favorable for all nanoparticles, but a non-monotonic trend in free energy barrier for adsorption is present as a function of linear alkyl end group length (and thus lipophilicity). End groups with low or high lipophilicity initiate adsorption through unfavorable lipid tail protrusions, whereas end groups of intermediate lipophilicity protrude favorably out of the ligand monolayer to intercalate into the lipid bilayer. We also show that ligand end group architecture (*i.e.*, number of end group branches) modulates the thermodynamics of adsorption. Ligand end groups of similar lipophilicity will exhibit lower free energy barriers as their degree of branching increases. We thus demonstrate that ligand lipophilicity and end group architecture are both important descriptors for nanoparticle adsorption to single-component bilayers. This work details a computational framework to study nanoparticle-membrane interactions that can be leveraged as a design space exploration tool for nanomaterials, and can potentially guide experimental frameworks in biomedical applications, such as drug delivery and biosensing.

[1] C. A. Huang-Zhu, J. K. Sheavly, A. K. Chew, S. J. Patel, and R. C. Van Lehn. "Ligand Lipophilicity Determines Molecular Mechanisms of Nanoparticle Adsorption to Lipid Bilayers. " *ACS Nano*, **2024**, *18(8)*, 6424-6437.

169q Investigating Phase Separation Behavior of Multi-Domain and Intrinsically Disordered Proteins with a Coarse-Grained Model.

Arjun Singh, Gregory Dignon

Phase separation is a fundamental phenomenon involved in cellular organization, and understanding the behavior of proteins undergoing phase separation is crucial. In this study, we develop a coarse grain model to investigate the phase separation behavior of intrinsically disordered proteins (IDPs) and multi-domain proteins, including folded domains. Our focus is on examining how the presence of folded domains influences the phase separation of these proteins. Our findings contribute to understanding the role of protein structure in phase separation and have implications for cellular processes. This also enables us to capture multi-component systems and understand the compositional control of membrane-less organelles more accurately.

169r Molecular Dynamics Simulations of Lipid Bilayer Mixtures: Developing Liposomes with Optimal Mechanical Properties

Stephen Adepoju, Francisco Hung

Liposomes and lipid nanoparticles are commonly used in delivery of drugs that have low stability, high hydrophobicity, or high cytotoxicity. Recent experimental results from the experimental group of Prof. Debra Auguste (ChemE, Northeastern) suggests that elastic liposomes have shown increased cellular uptake by cancer cells *in* vitro and increased tumor accumulation *in vivo*, when compared to more rigid liposomes. Using classical molecular dynamics simulations, we aim at fundamentally understanding how the mechanical properties of the liposomes are affected by variables such as the molecular structure of the lipids (tails, headgroups) and the composition of the liposome. The behavior of lipid bilayer mixtures consisting of 65% DOPC (1,2-dioleoyl-sn-glycero-3-phosphocholine) with 35% of either DSPS (1,2-distearyl-sn-glycero-3-phospho-L-serine) or DSPA (1,2-distearyl-sn-glycero-3-phospho-rac-glycerol) under various surface tension conditions was examined using molecular dynamics simulations with all-atom (CHARMM36) and coarse-grained (Martini) models. For every lipid mixture, we computed the area per lipid and the area compressibility modulus and compared against corresponding values in pure DOPC systems (considered to be the most elastic type of liposome). Excellent agreement was found when comparing some of our simulation results with available experimental data. Our findings show that the DOPC-DSPS system is stiffer than the DOPC-DSPA system, but both are softer than pure DOPC lipid bilayers. Insights about lipid membrane behavior, packing and mechanical characteristics of lipid bilayers under various surface tension settings will be presented and discussed. We will also investigate other properties such as the lipid tail order parameters, bilayer bending modulus and the lipid diffusion coefficients.

169s Dipole Moment Predictions Using Machine Learning Electron Density Models

Siddarth Achar, Leonardo Bernasconi, Karl Johnson

Accurate electron density information is essential for understanding chemical systems, particularly in scenarios involving dynamic processes like chemical reactions and ion transport. In materials chemistry, access to such data holds immense significance across various applications. The standard approach to predicting electron densities from quantum mechanics (QM) is limited due to high computational cost and poor scalability, constraining their application to smaller system sizes and shorter time scales for dynamics. To address this challenge, existing machine learning (ML) based electron density prediction methods, like DeepCDP (deep charge density prediction), provide a computationally efficient alternative for predicting QM-level electron densities with just the atomic coordinates as input. Useful physical properties like the dipole moment of a system can be computed just using electron density predictions. Efficient prediction of the dipole moment carries significant implications, for instance, in discerning whether proton transfer in a system follows proton-coupled electron transfer or hydrogen atom transfer. However, existing ML methods for electron density predictions face difficulties in predicting dipole moments with errors exceeding 10 Debye for some small molecules. In this study, we investigate the causes of errors in dipole moment prediction with ML models, utilizing the DeepCDP formalism. The original DeepCDP formalism employs the weighted smooth overlap of atomic positions to characterize environments on a grid-point basis and map them to electron density data generated from QM simulations. We curated a diverse selection of molecules spanning varying polarities, atomic radii, electronegativities, and types of bonds (covalent or ionic). Subsequently, we trained models for each of these systems independently and evaluated their predictions on electron densities and dipole moments. Our observations reveal that nearly all models predict electron densities with R2 values exceeding 0.99 and low mean squared errors. Models trained on nonpolar systems demonstrate better accuracy in predicting dipole moments compared to polar systems. We identify that systems containing atoms with larger nuclear charges exhibit the most significant prediction errors in dipole moments. This trend persists across both polar and nonpolar cases, with more pronounced errors observed in polar cases. Furthermore, we establish correlations between the nuclear charge in a system and the error in predicting the total number of electrons. These findings suggest that our current fingerprinting method struggles to correlate points at the center of atoms with high nuclear charge with their corresponding electron density, thus leading to errors in dipole moment prediction. We discuss avenues for parameter tuning and modifications to the atomic fingerprints to address these challenges.

169t Transferable Water Potentials Using Equivariant Machine Learning Interatomic Potentials

Tristan Maxson, Tibor Szilvasi

The ability of machine learning models to learn the underlying physics is critical for the transferability of models to different thermodynamic states(e.g., temperatures, phases, interfaces, etc.). Recent work[1,2] suggests that MLIPs trained on solely liquid-phase water (utilizing the MB-Pol potential as a ground truth) cannot accurately describe other phases such as the vapor-liquid equilibrium or gas-phase. An analysis of the many-body decomposition analysis of gas-phase water clusters shows that MLIPs rely greatly on error cancelation rather than reproducing fundamental interactions, limiting transferability. In this work, we demonstrate it is possible to train equivariant Allegro MLIPs on only 3,200 liquid-phase water structures and for the resulting MLIP to reproduce liquid-phase water properties (e.g., density within 0.003 g/cm3 between 230 and 365 K), vapor-liquid equilibrium density up to 550 K, and the relative energy and the vibrational density of states of ice phases. The many-body decomposition analysis of gas-phase water clusters up to six-body interactions shows that fundamental interactions are well described when equivariance is enabled (l=2), but results similar to the previous works are observed when we remove equivariance (l=0). The equivariant MLIP architecture appears to be capable of encoding complex descriptors which allow for improved transferability to arbitrary phases of water while remaining stable for nanoseconds to allow for full equilibration.

169u Perdew-Zunger Self-Interaction Correction for Ionization Energies of Transition Metal Atoms

Priyanka Bholanath Shukla, Rohan Maniar, Koblar Jackson, Karl Johnson, John P. Perdew

Density functional theory, commonly used for predicting chemical properties, suffers from self-interaction errors. These errors may result in errors of ionization energies and inaccurate descriptions of various other properties such as reaction barriers, molecular dissociation energies, charge transfer energies, and others. In this work, we employ the Fermi-Löwdin Self-Interaction Correction (FLOSIC) method to implement the Perdew-Zunger self-interaction correction (PZSIC), which is implemented on an orbital-by-orbital basis. Our goal is to assess the performance of PZSIC and locally scaled SIC (LSIC) in predicting the ionization energies of the transition metal atoms from changes of total energy. We compared the performance of uncorrected density functional approximations (DFAs) such as LDA, PBE and r^2SCAN against the reference experimental values for ionization energies. We also compared the results of PZSIC-LDA, PZSIC-PBE, PZSIC-r^2SCAN and LSIC-LDA with reference experimental values. We found that LDA and PBE overestimated the first three ionization energies of the Cu atom ($3d^{10}4s^1$), whereas r^2SCAN accurately predicted the ionization energies of Cu. PZSIC-LDA accurately predicted the first and third ionization energies but underpredicted the second ionization energy by approximately 2 eV. PZSIC-PBE and PZSIC-r^2SCAN further underestimated the second ionization energy by about 3 eV. To understand these errors in SIC methods, we hypothesized that there is a SIC energy penalty for a full 3d subshell in the PZSIC total energy for transition-metal atoms and ions. This penalty is not observed in the first ionization energy of Cu, because it cancels out when taking the difference between Cu and Cu^+. We observed this penalty when comparing Cu^+ and Cu^{+2} PZSIC total energies. We found similar energy penalties for atoms with $3d^5$ configurations such as Cr and Mn. Thus, there is an energy penalty when an electron is removed from a $3d^5$ or $3d^{10}$ shell of Cr, Mn, Cu, Zn, Ga, and Ge. On average, LDA, PBE and r^2SCAN overestimate the removal of the first 3d electron from a $3d^5$ or a $3d^{10}$ configuration by 1.32, 0.76, and 0.32 eV, respectively. PZSIC-LDA, LSIC-LDA, PZSIC-PBE and PZSIC-r^2SCAN underestimate the removal by 1.16, 0.11, 2.96, and 2.43 eV, respectively. LSIC-LDA gave the best performance among all uncorrected DFAs as well as PZSIC-DFA approaches. LSIC works by scaling down the size of the SIC correction, which reduces the energy penalty seen in $3d^5$ or $3d^{10}$ configurations. We suspect that the penalty in the PZSIC total energy is related to the nodal structure of local orbitals. For a filled 3d shell, the hybridization in FLOSIC leaves one or more of the orbitals less hybridized and more 3d-like, thus more noded. Our future work will focus on investigating the source of the SIC energy penalty in more detail and how to correct it within a PZSIC calculation.

169v Density Functional Theory (DFT) Analysis of CO_2 Adsorption and Dissociation on Reducible Oxides, and Integration into a Microkinetic Model for Dry Reforming of Methane (DRM)

Nirenjan Shenoy Padmanabha Naveen, Jonathan Lucas, Kerry Dooley, Michael Janik, Gina Noh, Konstantinos Alexopoulos

Mitigating greenhouse gas (GHG) emissions is imperative, and dry reforming of methane (DRM) is a promising avenue for upcycling two of the most potent GHGs. CH_4 and CO_2 are converted into syngas (H_2 and CO), which can be used as the feed for Fischer-Tropsch processes. Metal-support catalysts are usually employed for this reaction. CO_2 activation is one of the rate-determining steps and the activation mechanism is largely dependent on the nature of the support. Redox-active supports, such as ceria, exhibit strong favorability to form oxygen vacancies, which serve as active sites for CO_2 activation and dissociation. Whereas irreducible oxides like alumina follow the H-assisted CO_2 activation pathway, through the formation of CHO*/COOH* intermediates. However, the understanding of CO_2 activation/dissociation mechanisms on mixed oxide supports remains relatively limited.

Our experimental collaborators have synthesized a complex Ni-CeO_x-Al_2O_3 catalyst through atomic layer deposition of thin alumina layers on ceria. The incorporation of alumina overlayer suppresses the reverse water gas shift reaction, consequently enhancing the H_2:CO ratio. Meanwhile, ceria oxidizes the surface carbon species by supplying oxygen to the metal, while also facilitating CO_2 activation. Characterization techniques suggest that ceria and alumina exist as a mixed oxide in the catalyst. This observation is supported by our DFT calculations, which have elucidated the formation of a $CeAlO_x$ type surface oxide, with the Ce^{3+} species involved in CO_2 activation. These unique properties of the mixed-oxide support confer exceptional stability and selectivity to the catalyst. Preliminary investigation of the thermodynamic energetics of CO_2 activation and dissociation on the ceria-alumina support indicates that this mixed-oxide exhibits superior CO_2 activation capabilities compared to the respective pristine oxides.

To further understand the mechanism of CO_2 activation on the support, and influence of its kinetics on the overall catalyst performance, we have developed a dual site microkinetic model (MKM) based on elementary reaction rate constants obtained from DFT calculations. Through the MKM, we evaluate the rates of CO_2 activation/dissociation on the pure oxide supports vs the Ce-Al mixed oxide support. The reducibility of such mixed oxides determines the relative rates of H_2/CO formation, thereby influencing the catalyst activity. The degree of rate control analysis confirms the CO_2 activation on vacancies to be a rate-determining step, underscoring its role in overall reaction kinetics. Altogether, this work provides novel mechanistic understanding of the influence of CO_2 activation kinetics on the overall performance of the Ni-ceria-alumina catalyst.

169w Modeling the Effect of Surface Tension and Neutral Lipid Mixtures on the Structure of Lipid Monolayer Interfaces

Oluwatoyin Campbell, Viviana Monje-Galvan

In mammalian cells, lipid monolayers support the integrity of lipid droplets (LDs), organelles that function as storage for neutral lipids. Liver-targeting illnesses such as hepatitis C, non-alcoholic fatty liver disease and liver cancer interrupt normal lipid droplet metabolism and prompt changes in their chemical, structural and mechanical properties. In recent studies, the structure of LDs has been refined to include concentric crystalline phases of cholesteryl esters (CEs) and triglycerides near the interface where the neutral lipid core meets the lipid monolayer. This packing becomes more pronounced as CE concentration increases, showing a dependence of these structures on neutral lipid mixture. However, there is still need for molecular-level understanding of how lipid content and dynamics facilitate the presence of these structures within LDs. In this work, we use molecular dynamics simulations to model LD systems containing lipid mixtures of increasing CE mass% and evaluate changes in lateral organization and biophysical properties of the neutral core and phospholipid monolayer. To probe mechanical properties of the modelled LDs, we also observe changes under increasing surface tension. The results suggest a competition between (i) decreased ordered packing of the monolayer due to lateral stretching, and (ii) increased lipid packing due to insertion of fatty tails of CE into the monolayer. Stretching prompts vertical compression of the neutral LD core and increased repulsion between molecules at the monolayer-core interface. In addition, the tail end of the CE molecules prefers to localize in the monolayer as CE interdigitation equilibrates. To build on this work, the biophysical and mechanical effects of other key lipid species of LDs will be examined. Taken together with results from our collaborators, our insights improve understanding of the effects of liver diseases on the molecular arrangement and stability of LD monolayer-core interfaces.

169x Generalizability of Machine Learning Derived Interatomic Potentials of Peptides and Isomeric Structures

Lejla Biberic, Nisarg Joshi, Jim Pfaendtner

Biomolecular simulations are typically performed with classical force fields or using ab initio molecular dynamics (AIMD). Accuracy at the electronic level is lost in the former, while computational time limits the scalability of the latter.[1,2] Machine learning potentials (MLPs) trained on AIMD data offer higher accuracy results while overcoming the computational limitations of AIMD.[3] In this study we explore the generalizability of MLPs trained on AIMD data obtained from the density functional theory (DFT) version of Car-Parrinello Molecular Dynamics for alanine dipeptide and its peptoid isomer disarcosine. We demonstrate the computational speed up and ability of these MLPs to capture DFT level accurate simulations with a limited training set. We illustrate the versatility of MLPs to simulate peptide and peptoid systems. These results demonstrate how with a limited dataset a sufficient MLP can be developed to capture DFT level accuracy and generalizability across different simulation parameters of interest like temperature. Furthermore we demonstrate the capability of MLPs to capture rare events of the systems and obtain their free energy profile. This work contributes to the production of faster simulations with increased accuracy and generalizability and can be used to extend the types of biomolecules studied.

References:

1. D. Marx and J. Hutter, *Ab initio Molecular Dynamics: Basic Theory and Advanced Methods* (Cambridge University Press, 2009).
2. Jörg Behler; Perspective: Machine learning potentials for atomistic simulations. *J. Chem. Phys.* 7 November 2016; 145 (17): 170901.
3. Chmiela, S.; Sauceda, H. E.; Müller, K.-R.; Tkatchenko, A.Towards Exact Molecular Dynamics Simulations With Machine-learned Force Fields.Nat. Commun.2018,9, 3887.

169y Li Ion Diffusion in Solid Electrolyte Analyzed Using Deep Generative Models: Dependence of Accuracy of Diffusion Coefficients on MD Data Length.

Teppei Fukuya, Takayuki Nishio, Hiroya Nitta, Taku Ozawa, Kenji Yasuoka

Molecular Dynamics (MD) are effective in providing insights at the atomic level, However, the computational costs are notably high, particularly ab-initio MD (AIMD). Even as parallelization techniques have expanded the spatial scale of MD, the extension of the temporal scale remains a persisting challenge.

New computational approaches, particularly those incorporating machine learning like MD-GAN [1,2,3], are gaining traction for accelerating computation. MD-GAN, based on Generative Adversarial Networks (GAN), has the potential to generate long-term data from short-term data, thus enabling faster computation through temporal parallelization. To minimize computational costs, it is essential to keep the duration of MD calculations, utilized as training data, as short as possible. Nevertheless, excessively short MD calculations may result in an adequate number of samples, thereby impeding the attainment of the requisite accuracy for estimating physical quantities through training. In the previous study [4], we investigated the accuracy and optimal conditions of MD-GAN when applied to Li ion diffusion systems obtained using AIMD with SIESTA code [5]. In this presentation, we examine the effects on accuracy as the length of input data is shortened, aiming to determine the optimal number of MD calculation steps for MD-GAN training.

1. Endo, Tomobe, and K. Yasuoka, Proceedings of the AAAI Conference on Artificial Intelligence, 32, 2192 (2018).
2. Kawada, R., Endo, K., Yasuoka, K., Kojima, H. and Matubayasi, N., Journal of Chemical Information and Modeling, 63, 76-86(2022).
3. Kawada, R., Endo, K., Yuhara, D. and Yasuoka, K, Soft Matter, 18, 8446-8455(2022).
4. Fukuya, H. Nitta T. Ozawa, and K. Yasuoka , AIChE 2023 Annual Meeting

169z Exploration of PHA Synthase Mechanism through QM/MM Simulations

Youngbi Kim, Jeong Woo Han

Plastic waste accumulation is exacerbating global ecosystem destruction and pollution. Despite the continuous increase in plastic production, a significant portion of plastic remains unrecycled and ends up being incinerated, landfilled, or improperly managed. Consequently, there is a growing demand for biodegradable plastics to address the regulation of non-biodegradable single-use plastic products and the escalating plastic waste crisis. Among these biodegradable plastics, Polyhydroxy alkanoate (PHA) stands out as a promising solution. Despite its advantages of adjustable decomposition periods and cost-effective raw materials, PHA production suffers from inadequate productivity. However, PHA is considered the most ideal biodegradable plastic due to its unique ability to decompose not only in soil but also in seawater, making it the only biodegradable material capable of such widespread decomposition. Therefore, synthesizing PHA is crucial, requiring optimal pretreatment and bacterial enzyme optimization, especially to avoid bacterial carbon source excess, hindering synthesis. The final enzyme used in PHA synthesis relies on a specific substance, 3-hydroxybutyryl-CoA (3HB-CoA). Currently, the catalytic site of PHA synthase is known as HIS-CYS-ASP, but the mechanism remains poorly understood. In this study, we provide an atomistic and thermodynamic interpretation of the catalytic reaction mechanism of PHA synthase using umbrella sampling simulations at the robust B3LYP/MM MD level with a large QM region. The reaction mechanism involves proton transfer from Cys118 to His307, concerted with a nucleophilic attack by Cys118 on the substrate, leading to the release of CoA. Subsequently, another 3HB-CoA is inserted into the active site, and the elongation of 3HB occurs through the same process. Additionally, we utilized Density Functional Theory (DFT) to calculate the mechanism solely for the active site and compared the results with QM/MM simulation outcomes. This work enhances our understanding of the catalytic mechanism of PHA synthase and paves the way for further rational enzyme engineering efforts.

169aa A High-Throughput Screening Study to Design Ultra-High Performance Aramid Copolymers

Hyeonsuk Yoo, Ruth Muthoka, Yongjin Lee

Aramid fibers are a class of synthetic fibers characterized by exceptional properties, including high strength, excellent thermal resistance, and outstanding chemical inertness. These properties make them ideal for advanced composite and high-performance applications. However, engineering high-performance materials based on aramid fibers still faces challenges, particularly the chemical inertness of aramid fibers, which complicates processability into materials for specific applications and increases manufacturing costs.

In this poster, we propose a high-throughput screening approach to overcome these challenges and identify promising and novel aramid monomers for constructing advanced aramid chain structures with superior properties. Our screening methodology consists of two steps: 1) a cheminformatic screening based on molecular features, utilizing the extensive PubChem database, one of the largest databases containing unique chemical structures and property information, and 2) a molecular dynamics (MD) simulation-based screening, in which each monomer is evaluated for interchain interaction energies and subsequently tested via tensile strength simulations. Our screening approach successfully identified promising and novel monomers for aramid copolymers.

This research contributes to the ongoing advancement of polymer chemistry and the design of new materials, paving the way for developing next-generation aramid fibers with improved properties and cost-effectiveness.

169ab Integrating Biophysical Modeling and Machine Learning to Discover Plastic-Binding Peptides for Microplastic Remediation

Michael Bergman, Carol Hall

Plastics on the micrometer or nanometer scale, termed microplastics or nanoplastics (MNPs), are a harmful pollutant. Millions of tons of MNP pollution pervade the environment. Humans and many other organisms regularly consume MNPs unknowingly, which can have harmful effects on their health. Because the risks of MNP to the environment and human health worsen with increasing MNP concentration, we urgently need methods to remediate MNP pollution. Here we discover plastic-binding peptides (PBPs), which could detect and capture MNP pollution, by combining biophysical modeling and machine learning. PBPs were designed for polyethylene, polystyrene, polypropylene, and PET using the **Pep**tide **B**inding **D**esign (PepBD) algorithm. PepBD samples amino acid sequences and adsorbed conformations on a plastic surface using Metropolis Monte Carlo. Peptides were scored by the sum of their MMGBSA binding energy and peptide stability in the adsorbed conformation (i.e., the peptide internal energy). Molecular dynamics (MD) simulations of designed PBPs show that they consistently have stronger binding free energies than random amino acid sequences, while preliminary atomic force microscopy experiments found the best PBP for polyethylene has greater affinity than random sequences. An additional round of PBPs were designed to 1) increase PBP solubility in water to improve detection and capture of MNP pollution in aqueous environments, and 2) increase PBP binding affinity to a specific plastic to enable separation or characterization of the components of MNP waste. Water solubility was increased by adding the CamSol solubility term into the peptide scoring function. Binding specificity was improved by first training a long short-term memory (LSTM) network to predict the PepBD score of an amino acid sequence, and then searching for sequences with large score differences between plastics. Thorough experimental testing of PBP designs is ongoing. We envision that the designed PBPs will be broadly useful in addressing MNP pollution.

169ac Active Learning of Density Functionals with Error Control

Runtong Pan, Xinyi Fang, Mengyang Gu, Jianzhong Wu

Classical density functional theory (cDFT) offers a powerful statistical-mechanical approach for predicting the structure and thermodynamic properties of inhomogeneous fluids. However, practical applications of cDFT often face two fundamental challenges: 1) no systematic ways for the improvement of density functionals; 2) poor numerical scaling for solving the Euler-Lagrange equation. Functional learning by machine learning (ML) methods is likely to address both challenges. We have demonstrated that active learning with error control (ALEC) are able to build up a reliable emulator for predicting complex functionals with high computational efficiency by emulating cDFT calculations. In this work, we illustrate that active learning methods can also be used to establish machine-learning based density functionals with probabilistic algorithms. The active learning methods are more accurate than conventional ML models with "space-filling" data sampling such as random sampling and Latin hypercube sampling, and show both high computational and data efficiency.

169ad Customized Random Heteropolymer Design to Improve Protein Stability Using Molecular Dynamics Simulations

Yinhao Jia, Janani Sampath

Protein activity is known to decline outside their native environments, posing significant challenges in various fields. Polymers are promising materials used to either conjugate or mix with proteins to boost their function when subjected to changes in environmental factors. The recent development of random heteropolymers (RHPs) containing four distinct methacrylate-based monomers provides a promising new avenue of soft materials that possess the ability to shield proteins when mixed even without chemical conjugation, contributing to the improved activity and stability of proteins outside their native environment. While the controlled statistical composition design of RHPs has been proven to be effective, questions still remain regarding the ultimate efficiency of these conjugates, and whether designing custom polymers for different proteins might be a better alternative. Meanwhile, even with four types of monomers, the design space of RHPs is enormous. Given the broad design space of these materials, a mechanistic understanding of the stabilization would be beneficial for the rational design to fit different proteins.

In this study, we employ molecular dynamics simulations to assess the capacity of RHPs to improve lysozyme stability at high temperatures and understand the stabilization effect of the RHPs on proteins in the protein/RHPs mixture. Starting with a previously identified RHP composition – methyl methacrylate (MMA), oligo(ethylene glycol) methacrylate (OEGMA), 2-ethylhexyl methacrylate (EHMA), and 3-sulfopropyl methacrylate potassium salt (SPMA) in a 50:25:20:5 mole ratio and a polymerization degree of 80 – we generated simulated polymer sequences via the Composition Drift program. The stability of the protein in the mixture was evaluated using a high-temperature unfolding protocol. We then monitor the structural evolution of lysozyme at high temperatures. A promising stabilization effect was observed for lysozyme with a relatively stable and long-lasting secondary structure. Further analysis of the contacts between the protein residues and RHP monomers gives insight into the stabilization mechanism, enabling us to strategically design RHPs. Leveraging existing knowledge of protein's weak point, we further tailor RHP designs specifically for lysozyme, opening avenues for customized protein stabilization strategies.

169ae Molecular Dynamics Simulations of the Tear Film Lipid Layer to Elucidate the Causes of Dry Eye Syndrome

Mahbuba Khanom, Vance Jaeger

Dry Eye Syndrome (DES) affects nearly 20 million people in the United States, leading to a loss in quality of life due to constant irritation of the eyes. The most common cause of DES is defects in the tear film lipid layer (TFLL) which is a <100 nm thick layer of lipids that separates tears from the ambient air. The lipids in the TFLL originate in the meibomian glands in the eyelids. The TFLL is hypothesized to disperse the tear over the surface of the eye, to prevent airborne particles from entering the tears, and to prevent tear vaporization. When the TFLL is malformed due disfunctions of the meibomian glands, the TFLL no longer functions as intended, and patients experience a dry sensation in the eyes. The composition, properties, and precise balance of the TFLL structure directly leads to DES. Thus, understanding the structural and functional attributes of TFLL is essential for grasping the functioning of the tear film under both normal and pathological circumstances. While the chemical composition of the TFLL is well-established, the layer-by-layer structure of the TFLL is unknown. The structuring of these layers is key to developing DES drugs and therapies. Experimental methods for probing the TFLL are limited due to ethical considerations of experimenting on living eyes. Therefore, we have used molecular dynamics (MD) simulations to test hypothetical TFLL structural arrangements and support or refute prevailing published hypotheses.

Structures of 100 different lipids from the lipid classes such as wax esters, cholesterol esters, OAHFAs, phospholipids, sphingolipids, and glycerides. CHARMM36m and the CHARMM General Force Field was used to model common and uncommon lipids. Lipids were packed together using Packmol to follow hypothetical TFLL structures. Five different arrangements of the TFLL were constructed, and the models contained nearly 1 million atoms. MD simulations of at least 1 μs were conducted to allow for structural relaxation and to provide estimations of properties that have been measured experimentally. Simulations were thus validated against experimental measurements of surface tension, viscosity, rotamer distribution, and lipid crystallinity. Based upon these simulations, some of the prevailing hypotheses have been refuted, leading to a reduced set of plausible TFLL structures. These findings provide plausible atomistic structures that allow for future *in silico* testing of drug molecules that can modify the properties of the TFLL, thereby restoring function and providing relief to millions of patients.

169af Modeling H-D Exchange in Supported Catalytically Active Liquid Metal Solutions Using Reactive Molecular Dynamics

Gunnar Sly, Kathryn MacIntosh, Adri C.T van Duin, Robert Rioux, Michael Janik

Supported catalytically active liquid metal solutions (SCALMS) represent a unique class of materials owing to the dynamic nature of late transition metals introduced to a liquid metal matrix. SCALMS' bi-functional nature can enable propane dehydrogenation reactions while achieving exceptional H_2 absorption capacity. These intrinsic properties allow for process intensification via simultaneous catalytic conversion and reactant separation. However, the atomic structure of SCALMS, specifically PdGa, and the interactions with gaseous H_2 and relevant C_3 species (propane/propylene) is challenging to probe experimentally. Density functional theory (DFT) allows for the quantum mechanical description of short-range materials but is incapable of accurately describing liquid metal systems due to the dynamically unstable configurations of the liquid solution. Reactive molecular dynamics, using the ReaxFF reactive force field, facilitates the study of these dynamic systems as well as transport and kinetic phenomena directly observed over time.

We present a ReaxFF-based investigation of the Pd-Ga-C-H system to understand the ensemble average Pd interactions with molecular or atomic hydrogen within the liquid matrix as well as induced surface responses under the presence of gaseous alkanes. To ensure our model describes both the representative structures and kinetics of experimentally synthesizable SCALMS, we evaluate H_2 permeability through the liquid and the equilibrium H_2 absorption as a function of Pd content. To achieve thermodynamically stable PdGaH phases, a Grand-Canonical Monte Carlo (GCMC) methodology is applied which permits probing H-loading at various temperatures, H_2 pressures, and Pd contents for experimental validation. Additionally, we introduce into our simulations a mixed propane/propylene gas phase to study adsorption-induced surface segregation of Pd to the surface of the liquid matrix which has important implications towards the H_2 dissociative mass-activity of the late-transition metals within SCALMS.

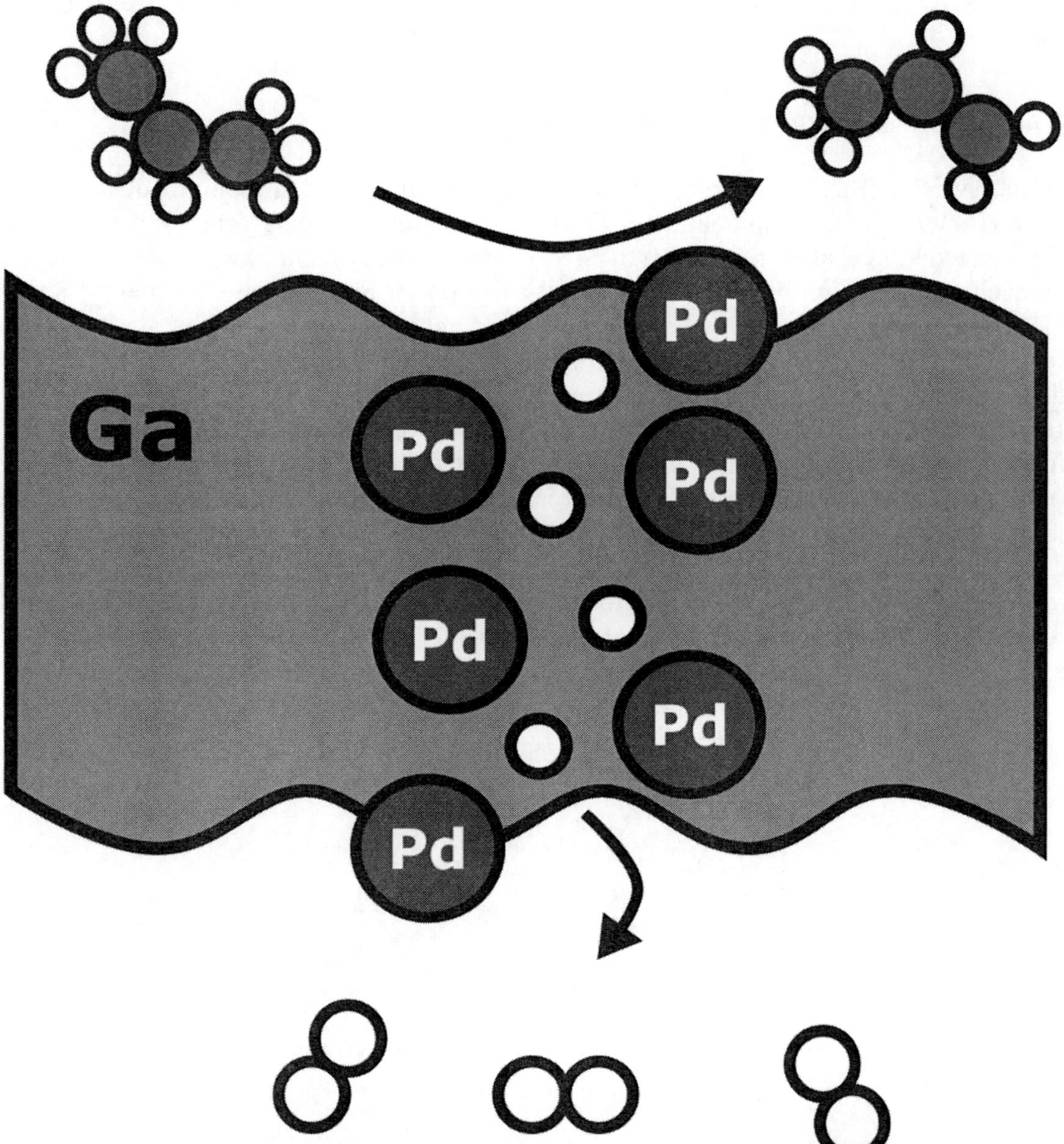
Ga
Pd
Pd
Pd
Pd
Pd
Pd

169ag Enhanced Shockwave Synthesis through Accurate Silicon Modeling Using Chimes

Thomas Sundberg, Safa Alzaim, Rebecca Lindsey

Silicon nanoparticles (SiNP) and composites like silicon carbide and silicon nitride are of technological interest in the semiconductor domain. However, a precise yet high throughput synthesis capability is needed to enable integration of SiNPs into everyday technological devices. Existing synthesis techniques are inhibited by either low throughput or limited material quality. Shockwave synthesis techniques, such as explosive deposition, can achieve both the production rate and quality required, but the extreme conditions reached, which can reach 7000K and 90 GPa have thus far precluded establishing a clear picture of the underlying mechanisms and kinetics needed to enable finely tunable SiNP synthesis through experiments alone. Atomistic simulations offer a complimentary route to studying this process but rely on availability of "quantum-accurate" interatomic models capable of faithfully describing the complicated evolution that ensues under extreme conditions. Therefore, in this presentation, we discuss recent efforts to develop a machine learned interatomic potential (ML-IAP) for silicon model using ChIMES, a physics-informed framework ML-IAP designed specifically to enable large scale yet quantum-accurate simulation of materials under extreme conditions. Ultimately, this model can serve as a foundation upon which models for other silicon-containing materials (e.g., Si-C) can be generated, enabling simulation-guided refinement of high pressure (e.g. shock-) synthesis strategies for Si-based nanomaterials.

169ah Uncovering Residue-Level Driving Forces Underlying the Formation of Biomolecular Condensates

Shiv Rekhi, Jeetain Mittal

The self-assembly of intrinsically disordered proteins (IDPs) into membraneless compartments, termed membraneless organelles (MLOs) or biomolecular condensates, has implications in understanding cellular function, as well as the design novel biomaterials. Despite the interest in MLOs, a fundamental understanding of the thermodynamic driving forces linking amino acid sequence to phase behavior remains elusive. It has been hypothesized that certain sequence features, specifically, the presence of aromatic residues, and charged residues particularly Arg, encode the phase behavior of prion-like IDPs. Here,[1] we target previously overlooked sequence features through the design of sequence variants of a model polypeptide to investigate if, and how, amino acids and sequence features that fall outside of the existing molecular grammar alter phase behavior. We find that all but one of the sequences designed undergo phase separation, pointing to a model of all residues collectively contributing to the driving forces for phase separation, rather than just a handful. Further, we find that the impact of residue substitutions on the driving forces for phase separation are not explained by their solvation thermodynamics. To identify the cause for the deviations from bulk-phase hydrophobicity measures, we perform alchemical free energy calculations to quantify the free energy of transfer of amino acid sidechain analogs from the water-rich dilute phase to the protein-rich dense phase. We observe a complex interplay between protein-sidechain interactions within the condensate, and sidechain-solvent interactions within the dilute and condensed phases in determining the net favorability or unfavorability of transfer free energies of all amino acids. To summarize, in this work we uncover the contributions of residues previously overlooked in the molecular grammar of phase separation, as well as provide a molecular picture of the forces underlying their contribution to formation and stabilization of biomolecular condensates.

References

1. Rekhi, Shiv, et al. 'Expanding the molecular language of protein liquid–liquid phase separation.' *Nature Chemistry* (2024): 1-12.

169ai A Synergy of Molecular Simulation, Mathematical Programming, and Machine Learning for the Phaseout of Harmful Refrigerants

Barnabas Agbodekhe, Edward Maginn

In 2016, 197 countries signed the Kigali agreement to phase out refrigerant fluids with high global warming potential (GWP) that can cause adverse climate change effects. These refrigerant fluids are often azeotropic or near-azeotropic mixtures of two or more hydrofluorocarbons (HFCs). One of the major technical hurdles that must be surmounted to enable phaseout requirements is the separation of these high GWP refrigerant mixtures, for reuse and recycling, to prevent them from being vented or incinerated. The other critical need is discovering and designing new, greener refrigerant fluids. We have applied several computational molecular science tools synergistically to help address these challenges.

Addressing the first challenge requires the design of advanced separation technologies. Extractive distillation using ionic liquids (ILs) as entrainers has been proposed as a viable solution. However, there is limited understanding of IL/HFC systems, including transport and interfacial data. We applied molecular dynamics (MD) simulations using several advanced methods and techniques for reliable studies of thermal conductivity, viscosity, self-diffusivity, and interfacial tension for several IL/HFC systems. Our results were able to excellently capture experimentally observed qualitative trends for these properties for all IL/HFC systems studied. We also investigated and successfully elucidated the molecular origins of some of these properties and the link between the molecular liquid structure and these properties. Our data and insights from MD simulations will support secondary-level screening of ILs to select or design the 'optimal' IL for use in separating HFCs.

To address the second challenge, we applied mathematical programming in a molecule generation tool we developed called FineSMILES to exhaustively generate all or almost all acyclic molecules that could serve as refrigerants in vapor-compression refrigeration cycles, based on thermodynamic arguments. We generated hundreds of thousands of molecules, more than 50 % of which were not reported in the PubChem database. We then applied machine learning for property prediction to rapidly screen these generated molecules based on technical, safety, and environmental performance considerations. We used the retrosynthetic accessibility scores for the preliminary assessment of synthesizability. We obtained several tens of computationally discovered molecules that have high potential for use as refrigerants and may be likely synthesizable. Finally, we further applied high throughput molecular simulations to assess the technical performance of the discovered fluids and to design and assess azeotropic or near-azeotropic green refrigerant mixtures using the discovered molecules as components. Five novel, high-performing green refrigerant mixtures were designed and recommended for further experimental testing.

169am Evaluation of Effective Dielectric Constants: Implications for Modeling Protein-DNA Liquid-Liquid Phase Separation

Vithurshan Varenthirarajah, Utkarsh Kapoor, PhD

Electrostatic interactions are a fundamental force governing the behavior of biological macromolecules such as proteins and nucleic acids. Therefore, accounting for electrostatic interactions correctly is crucial for understanding mechanisms of biological processes. In general, the existing CG protein/DNA models use the implicit solvent potential based on Debye-Hückel theory, where dielectric permittivity corresponds to water ($D \approx 80$ at 300K). However, in the vicinity of the protein/DNA, this dielectric is a factor of $\approx 2 - 10$ too large and thus protein-DNA interactions are underestimated. Further, protein-DNA interactions are known to be mediated by water, a phenomenon excluded from the models by the use of implicit solvent (i.e. Langevin dynamics). This suggests the need to account for spatial variation. Seeking to use the existing protein/DNA CG models with minimal modification, this research project aims to evaluate the impact of effective dielectric constants by incorporating a distance, temperature, or concentration-dependent dielectric constants and provide recommendations for the appropriate dielectric constants for the calculation of electrostatic interactions. Using prototypical protein-DNA systems that undergo liquid-liquid phase separation (LLPS), we access the validity of these models by comparing their predictive capability of reproducing local structure properties and binding affinities. Overall, this research holds promise and can be used to investigate protein–DNA complexes, enabling a mechanistic understanding of various biological and engineering applications.

169an Force Field Development for Pyrrolidinium-Based Ionic Liquids for Ionic Conductivity Predictions

Ashutosh Kumar Verma, Amey Thorat, Jindal Shah

Ionic liquids (ILs) are becoming increasingly popular as potential candidates for the next generation of energy storage devices due to their low melting points, low volatility, non-flammability, exceptional stability, and broad electrochemical potential window. Ionic liquids are ideal electrolytes for the electrochemical and energy storage devices. Given the large number of ionic liquids that can be designed, it is necessary to screen ionic liquids to identify the ones with high ionic conductivity. Molecular simulation-based approaches have been applied over the last couple of decades for ionic liquid screening. However, for estimating transport properties such ionic conductivity, polarization effects due to the bulk environment need to be incorporated into force fields employed in molecular simulations. In this poster presentation, we show how atomic partial charges are calculated for ionic liquids based on the 1-butyl-1-methylpyrrolidinum cation paired with dicyanamide, bis(trifluoromethanefulonyl)imide, and trifluoromethanesulfonate anions. These charges are coupled with the bonded and non-bonded Lennard-Jones parameters obtained from Generalized Amber Force Field (GAFF) and Optimized Potentials for Liquid Simulations (OPLS). Results of density, ionic conductivity, and ion-ion correlations contributing to the ionic conductivity will be presented.

169ao Permutationally Invariant Network for Enhanced Sampling (PINES): A General Approach to Treating Identical Particles and Constructing Targeted CVs with Machine Learning

Nicholas Herringer, Siva Dasetty, Diya Gandhi, Junhee Lee, Andrew Ferguson

Collective variable (CV) based enhanced sampling techniques are commonly used to address the presence of high free energy barriers in molecular simulation and to subsequently improve sampling of phase space. Data-driven methods present a powerful means to identify high-quality CVs[1]. It is often desirable to learn CVs respecting the underlying symmetries of the molecular system such as translational, rotational, and permutational invariances. Determination of CVs invariant or equivariant to translation and/or rotation can be readily achieved by data preprocessing or the use of specialized learning architectures[2], but it has proven more challenging to engage the permutational invariance associated with the atomic or molecular indistinguishability[3]. We introduce Permutationally Invariant Networks for Enhanced Sampling (PINES)[4] as a data driven method for learning translationally, rotationally, and permutationally adapted CVs and demonstrate its application to a series of test systems, including Lenard-Jones clusters, ion association in water, and hydrophobic polymer collapse. Here we present recent innovations in our code to generalize PINES to arbitrary systems and detail a systematic user guide to application through the enhanced sampling software suites PLUMED2[5] and SSAGES[6].

References:

(1) Sidky, H.; Chen, W.; Ferguson, A. L. Machine Learning for Collective Variable Discovery and Enhanced Sampling in Biomolecular Simulation. *Molecular Physics* **2020**, *118* (5), e1737742. https://doi.org/10.1080/00268976.2020.1737742.

(2) Atz, K.; Grisoni, F.; Schneider, G. Geometric Deep Learning on Molecular Representations. arXiv December 31, 2021. http://arxiv.org/abs/2107.12375.

(3) Noé, F.; Tkatchenko, A.; Müller, K.-R.; Clementi, C. Machine Learning for Molecular Simulation. *Annual Review of Physical Chemistry* **2020**, *71* (1), 361–390. https://doi.org/10.1146/annurev-physchem-042018-052331.

(4) Herringer, N. S. M.; Dasetty, S.; Gandhi, D.; Lee, J.; Ferguson, A. L. Permutationally Invariant Networks for Enhanced Sampling (PINES): Discovery of Multi-Molecular and Solvent-Inclusive Collective Variables. arXiv August 16, 2023. https://doi.org/10.48550/arXiv.2308.08680.

(5) Tribello, G. A.; Bonomi, M.; Branduardi, D.; Camilloni, C.; Bussi, G. PLUMED 2: New Feathers for an Old Bird. *Computer Physics Communications* **2014**, *185* (2), 604–613. https://doi.org/10.1016/j.cpc.2013.09.018.

(6) Sidky, H.; Colón, Y. J.; Helfferich, J.; Sikora, B. J.; Bezik, C.; Chu, W.; Giberti, F.; Guo, A. Z.; Jiang, X.; Lequieu, J.; Li, J.; Moller, J.; Quevillon, M. J.; Rahimi, M.; Ramezani-Dakhel, H.; Rathee, V. S.; Reid, D. R.; Sevgen, E.; Thapar, V.; Webb, M. A.; Whitmer, J. K.; de Pablo, J. J. SSAGES: Software Suite for Advanced General Ensemble Simulations. *The Journal of Chemical Physics* **2018**, *148* (4), 044104. https://doi.org/10.1063/1.5008853.

169ap Atomic-Level Structural Model of Helical Tdp-43 Oligomers

Azamat Rizuan, Jayakrishna Shenoy, Priyesh Mohanty, Nicolas L. Fawzi, Jeetain Mittal

TAR DNA binding protein-43 (TDP-43) is a key nuclear RNA-binding protein implicated in RNA metabolism such as transcription and splicing. The C-terminal domain (CTD) of TDP-43 is intrinsically disordered and aggregation-prone, commonly found in cytoplasmic inclusions associated with neurodegenerative disorders including amyotrophic lateral sclerosis (ALS) and frontotemporal dementia (FTD). CTD contains a conserved helical region (CR, spanning residues 319-341) which adopts a transient α-helix structure and drives the formation of higher-order oligomers which are important for its splicing and nuclear retention. Notably, ALS-associated mutations in the CR disrupt oligomerization and function, further highlighting the importance of the CTD in disease. Therefore, uncovering the atomistic details underlying the formation of helix-mediated CTD oligomers is crucial for the development of novel therapeutic strategies for ALS/FTD. However, achieving a high-resolution, structural characterization of these high-order oligomers remains challenging, mainly due to CTD's high aggregation propensity and dynamic assembly. To overcome these challenges, we employed an integrative modeling strategy based on state-of-the-art computational methods including AlphaFold2-Multimer modeling and all-atom molecular dynamics (AAMD) simulations, along with experimental insights from biophysical experiments, to probe the atomic structural details of TDP-43 helical self-assembly. We first characterized the structural ensemble of the TDP-43 CR dimer state (aggregate time ~100 μs), demonstrating the essential role of helix-helix contacts in driving the formation of dynamic TDP-43 CTD higher-order assemblies. Our simulations suggest that CR dimers lack a stable orientation and instead exhibit dynamically interconverting structures with heterogeneous helix-helix interactions. Through integrative modeling, we identified a uniquely stable atomic structural model of the TDP-43 CR tetrameric state, in qualitative agreement with experimental measurements from NMR spectroscopy. Importantly, our structural model elucidates, for the first time, the crucial structural role of the G335-Q343 region, a hotspot of ALS mutations, which becomes partially helical only upon higher-order assembly. Overall, our study makes significant advances towards structural characterization of TDP-43 functional assemblies and provides clues regarding their transition into pathogenic aggregates.

169ar Capturing Realistic Double Layer Cation Properties in Classical MD Using Aimd-Guided Potential Fitting

Derek Zhu, Vitaly Alexandrov, Michael Janik, Scott T. Milner

Understanding electrochemical processes is fundamental for advancing various technological applications, from energy storage to catalysis. The distribution and behavior of cations within the electric double layer (EDL) play a crucial role in these processes. Ab initio molecular dynamics (AIMD) simulations offer a powerful tool for studying these phenomena with accurate electronic structure analysis. However, despite their strengths, AIMD simulations also come with inherent limitations, particularly when investigating cation distributions throughout the entire double layer. The high computational cost of these quantum mechanical (QM) methods, combined with the slow dynamics of ions and water when diffusing to and from the electrode surface, severely hinders the usefulness of AIMD when exploring ion distributions in the EDL. Classical models, on the other hand, provide an alternative approach that can efficiently capture the dynamic behavior of cations over longer timescales and larger system sizes. We have developed a dynamic charge approach ("QDyn"), which takes advantage of classical electrostatic principles, to rapidly allow charge in the (non-neutral) metal to move in response to electrolyte dynamics. This method allows simulation of 10's of nanoseconds of dynamics for systems containing 10,000's of metal and electrolyte atoms over a few days on a modest lab computational resource. However, to ensure the reliability and accuracy of the model in capturing the behavior of ions within the EDL, it is essential to fit the metal ion potentials. Fitting these potentials to gas phase or micro solvated QM calculations poses challenges due to their limited accuracy in sampling many water configurations close to the system that would be present in the classical model. To refine the metal ion potentials, we used meta-dynamics biased AIMD to confine the ions close to the metal surface. This restraint was used to keep the ion close enough to the metal to measure the repulsive forces while improving sampling of the water solvation in this region. Fitting the behavior and repulsion of the ion meta-dynamics biased AIMD to a classical simulation with the same bias enabled a more accurate classical model which can then be used to better study the behavior of ions close to the interface. The repulsive trends of the ions are compared down the periodic table and across metals hinting at a systematic approach for estimating more accurate potentials without relying on individual AIMD simulation for every ion. Additionally, EDL properties of ions are studied by predicting the ion distribution and dielectric constant of the double layer as a function of ion concentration, identity, and metal potential. This comprehensive approach provides insights into the behavior of ions near the interface. Finally, recommendations on best practices for utilizing AIMD to fit classical potentials are presented, offering guidance for future research utilizing classical molecular dynamics.

169as Adsorbate Effects on the Mobility of Single Atom Catalysts: Using DFT to Map out Diffusion Potential Energy Surfaces

Shirin Asadialahsavand, Michael Janik, Konstantinos Alexopoulos

Representing a novel frontier in heterogeneous catalysis, Single-atom catalysts (SACs) supported on metal oxide surfaces exhibit remarkable potential for catalyzing hydrodeoxygenation reactions, boasting high activity and selectivity. However, the pronounced mobility of SACs renders them susceptible to aggregation, resulting in catalyst deactivation (Yang et al., 2013). Our study investigates the underlying mechanisms governing the aggregation behavior of SACs in the presence of atomic adsorbates. Employing density functional theory (DFT), we develop a comprehensive and explicit approach to enumerate the full Potential Energy Surface (PES), accounting for both metal atoms and adsorbates, to understand how adsorbates influence metal atom diffusivity. Our preliminary results unveil intriguing dynamics: while the adsorption of hydrogen on single metal atoms enhances surface mobility, the transition to more reduced species, characterized by hydrogen adsorption on the surface leading to hydroxyl group formation, introduces a physical and chemical barrier to diffusion, impeding metal atom mobility even in the hydride form. Furthermore, under elevated hydrogen pressures, increased hydrogen uptake triggers the formation of surface oxygen vacancies on the metal oxide. We propose that these vacancies act as nucleation sites, capturing single atoms and curtailing surface diffusion. Our findings illuminate the complex relationship between atomic adsorbates and SACs aggregation behavior, offering insights crucial for tailored design strategies to mitigate catalyst deactivation in hydrodeoxygenation processes.

Yang, X.-F., Wang, A., Qiao, B., Li, J., Liu, J., & Zhang, T. (2013). Single-Atom Catalysts: A New Frontier in Heterogeneous Catalysis. *Accounts of Chemical Research*, *46*(8), 1740–1748. https://doi.org/10.1021/ar300361m

169at Parametrization of Monatomic Ion-Biomolecular Interactions in the Polarizable Drude Force Field: Application in Protein and Nucleic Acid Systems

Yiling Nan, Alexander D. MacKerell

An accurate force field is the foundation of reliable results from Molecular dynamics (MD) simulations. The traditional additive force fields average the impact of atomic polarizability, which encounter difficulties in precisely capturing the electronic response across varied environments. As a result, these additive force fields often demonstrate limited reproducibility of thermodynamic properties observed in experimental conditions, especially in high ionic concentration systems and complex, heterogeneous conditions. This type of heterogeneity is prevalent in biological contexts, such as ion channels and nucleic acids, as well as in various industrial applications like Li^+ ion batteries and ionic liquid systems. The advancement in polarizable force fields, which explicitly considers atomic polarizability, can tackle this issue. In our recently published work, we developed a protocol to generate atom-pair specific LJ (known as NBFIX in CHARMM) and through-space Thole dipole screening (NBTHOLE) based on readily accessible quantum mechanical (QM) data to fit condensed phase experimental thermodynamic benchmarks (osmotic pressure, diffusion coefficient, ionic conductivity, and solvation free energy) when available. It is worth noting that condensed-phase experimental properties are often unavailable for most functional groups and ion complexes. This protocol proves particularly beneficial in generating parameters for those lacking experimental data. In the present work, we further qualify the developed protocol to generate the NBFIX and NBTHOLE parameters in the application of the interactions between monoatomic ions (specifically Li^+, Na^+, K^+, Rb^+, and Cs^+) and common functional groups found in proteins and nucleic acid systems. The parameters generated individually for each ion and functional group were allocated to their corresponding functional groups within proteins or nucleic acids. Subsequently, the distribution of ions around these functional groups in proteins and nucleic acids was examined. Our findings demonstrate that a modified force field successfully addresses the issue of over-binding observed in the previous iteration and accurately reproduces the effective charge of the protein. In addition, the simulation on nucleic acid in 100mM NaCl solution showed good agreement with the counterion condensation theory (CC) with DNA charge neutralized around 73 % (76% in CC theory) in the 1BNA system. This study aims to offer a more comprehensive and refined understanding of molecular interactions, extending beyond biological systems to encompass various fields across diverse scientific domains.

169av Tackling Energy Conversion Challenges through Simulations in Synergy with Experiments

Ritesh Kumar, Reginaldo Gomes, Hannah Fejzic, Chibueze Amanchukwu

The challenges posed by climate change have increased global interest in sustainable fuel production, particularly the advancement of electrochemical carbon dioxide or carbon monoxide reduction (CO2R or COR) technologies. In this regard, my research leverages the accuracy of molecular dynamics (MD) and density functional theory (DFT) simulations to augment experimental efforts in our group to enhance the selectivity and efficiency of CO2R and COR processes. By intricately analyzing the solvation dynamics and structural properties of a variety of aprotic solvents, we aim to systematically modulate the competing hydrogen evolution reaction (HER), often a significant barrier in both CO2R and COR efficiencies. The DFT and MD simulations offer atomistic insights into the solvation structure and dynamics within these solvents, enabling the identification of key descriptors that predict experimental HER and CO2R or COR selectivity. This approach allows for a targeted modification of solvent environments, guiding the design of electrolytes that favor CO_2 conversion to CO and CO to ethylene formation over HER even in the presence of water. The use of these sophisticated computational tools not only sheds light on fundamental mechanistic pathways but also paves the way for the rational design of more selective and efficient CO2R catalysts. Through this focused inquiry into the molecular underpinnings of electrolyte behavior, our work contributes to overcoming one of the significant challenges in the field of energy conversion, thereby supporting the transition towards a more sustainable and carbon-neutral energy landscape.

169aw Elucidating Gas Diffusion Dynamics at the MOF-Polymer Interface in Mixed-Matrix Membranes: A Computational Study

Moein Hajian Z.D., Hyunhee Lee, Zachary Smith, Joshua Howe

Our study focuses on a detailed computational investigation into the dynamics of gas diffusion across Mixed-Matrix Membranes (MMMs), where Metal-Organic Frameworks (MOFs), specifically Zeolitic Imidazolate Framework-8 (ZIF-8), are integrated with polymer matrices, such as 6FDA-DAM polymers. Utilizing techniques including Density Functional Theory (DFT) and Climbing Image Nudged Elastic Band (CI-NEB) calculations we meticulously analyze the electronic and structural characteristics of the interface between ZIF-8 and polymers. This analysis focuses on understanding the passage of various gases—hydrogen, nitrogen, methane, ethylene, ethane, propane, and propylene — through ZIF-8's most stable surfaces (100 and 110), in both their pristine and polymer-modified states.

A primary aim of this work is to identify the electronic interactions at the MOF-polymer interface and assess how these influence the MMM's efficacy for gas separations. By correlating our computational findings with experimental data on gas diffusion through MMMs, we underscore the significance of computational studies in understanding the complex interactions at the MOF-polymer interface. This effort is directed towards contributing insights that will aid in the advancement of more effective and environmentally friendly gas separation techniques.

169ax Using Deep Learning to Accelerate the Molecular Simulations and Predict the Kinetics of RNA Folding

Ayush Gupta, Heng Ma, Arvind Ramanathan, Gül Zerze

Understanding the process of RNA stem-loop folding and unfolding is crucial in biology. The poster will address the computational challenge of simulating RNA stem-loop folding due to the complex folding landscape that requires extensive computation. The research explores how to study the folding of RNA stem-loops using a deep learning-driven simulation method. We adapted DeepDriveMD (DDMD), a deep learning technique, to simulate RNA stem-loop folding dynamics. DDMD adaptively learns a low-dimensional latent representation from an ensemble of running MD simulations. DDMD then guides the simulations toward the undersampled regions while optimizing the resources to explore the relevant parts of the phase space. The method achieves reasonable free energy landscape prediction at room temperature and identifies relevant slow degrees of freedom in RNA folding. The kinetic rates of transition between the stable and metastable states were calculated using Markov state models. The qualitative analysis of the latent space indicated that the computational framework could capture the distinct conformations within the RNA folding process. DDMD can simulate RNA folding more efficiently than conventional methods, providing insights into phase space and system kinetics without extensive computational costs. This provides a feasible and more rapid approach for exploring the complex process of RNA folding, with broad implications for computational biology.

169ay Understanding Plasma-Driven Solution Electrochemistry

Subhajyoti Chaudhuri

Many applications involving plasma-liquid interactions depend on the reactive processes occurring at the plasma-liquid interface. In this poster, I discuss how a combination of classical and quantum mechanical methods can help us understand and predict these reactive processes.

169az Examining Ion Dehydration Mechanisms at the High Pressures and Concentrations Required for Desalination

Nathanael Schwindt, Anthony Straub, Razi Epsztein, Michael Shirts

High-salinity brine treatment is critical to increase water supplies and minimize waste discharge. Membrane-based desalination processes such as reverse osmosis (RO) and nanofiltration (NF) are being increasingly considered for brine treatment due to their high energy efficiency compared to thermal processes. However, brine treatment exposes membranes to extremely high pressures and salinities that alter membrane transport properties. The underlying mechanisms by which pressure and concentration affect the rejection of ions have not been fully determined. In this work, we study the effect of high pressures and salinities on the ion dehydration mechanism using atomistic molecular simulations of ions in aqueous solution and within an RO membrane. We explore these trends for ions of different size, valency, and charge. We quantify the energy barriers to remove one or more waters from the hydration shell for many relevant salts and operating conditions. We show that system pressure in the operating range of RO and NF membranes does not change the free energy of dehydration in solution. However, at high concentration, the free energy to strip waters is reduced. Stripped waters can more easily coordinate with other ions than disrupt the second hydration shell or bulk water structure. We observe the geometry of dehydration within an RO membrane in order to determine how confinement and membrane groups influence the hydration shell. Our results will provide a valuable reference for identifying the primary mechanisms for ion transport in industry-standard membrane materials.

169ba Physical Simulations of Genome Organization and the Histone Code

Soren Kyhl, Juan J. de Pablo

The three-dimensional structure of chromatin is increasingly understood to be a major player in determining gene regulation and genome function. Interest has risen in the prospect of therapeutics that are engineered to tailor the functional role of chromatin by modulating its' 3D structure. However, while

some new biochemical methods are available to precisely edit epigenetic marks in vivo, very little is known about the downstream effects of these edits on chromatin physical organization. We present a physical model for genome organization which represents epigenetic marks explicitly, enabling for the first time de novo epigenetic engineering in a predictive manner.

169bc Origin of the Stokes-Einstein Deviation in Liquid Al-Si

Ni Zhan, John Kitchin

One goal in materials simulation is finding relationships between atomic scale and continuum properties. The Stokes–Einstein–Sutherland (SES) equation relates diffusion and viscosity in liquids through the effective diameter d, which is frequently observed to be constant with the ratio $T/(D\eta)$ where T is the temperature, D is the diffusion coefficient, and η is the viscosity. The SES has a practical use in estimating diffusion from viscosity or vice versa as they can be difficult to measure experimentally. The constant effective diameter observed in the SES holds for many liquids within a temperature range, but at low temperatures, deviations where d is not constant have been observed. This work investigates SES in liquid Al–Si using molecular dynamics. We analyze the local order using Voronoi polyhedrons and agglomerative clustering. Clustering methods from machine learning allowed us to analyze the large amount of data generated from molecular dynamics trajectories in an efficient manner. We found that clusters have minimal effect on diffusion while increasing viscosity, which is a likely origin of the SES deviation for liquid Al–Si at low temperatures near the melting temperature.

169bd Exploring the Effects of Tributyltin Exposure on Adipocytes: An Integrated Transcriptomics and Metabolomics Study to Decode Molecular Response Linked to Metabolic Syndrome

Dayna Schultz, Ilias Frydas, Nafsika Papaioannou, Thanasis Papageorgiou, Catherine Gabriel, Spyros Karakitsios, Denis Sarigiannis

Endocrine disrupting chemicals (EDCs) are compounds that are known to disrupt physiological processes, such as hormonal regulation, which can in turn lead to adverse health outcomes in humans. EDCs comprise a significant number of environmental pollutants that are indiscriminately released into the environment every year, and most of which have endocrine-disrupting potential. Metabolic syndrome (MetS) is a cluster of seemingly vague medical conditions frequently occurring together and includes obesity, insulin resistance, dyslipidemia, and hypertension, and which have been associated with increased risk of cardiovascular disease and type 2 diabetes. Moreover, MetS is increasingly thought to be attributed, at least in part, to exposure to EDCs through inhalation, ingestion, or absorption from various exposures within the environment. Adipose tissue consists of various cell types such as adipocytes, pre-adipocytes, stromal vascular cells, extracellular matrix components, and blood vessels, and is a highly complex and dynamic tissue that plays a vital role in metabolic regulation. Adipocytes are the primary cell type that compose adipose tissue and is the major repository site for the storage and release of energy in the form of lipids, but also contribute to endocrine function and hormone sensing and response. Certain EDCs have shown the potential to interfere with normal development and function of adipose tissue. To better understand the effects of EDCs on metabolic syndrome, adipocytes were treated to a known lipogenic substance, tributyltin (TBT), and the -omic signature associated with metabolic dysregulation was mapped.

Simpson-Golabi-Behmel syndrome (SGBS) pre-adipocytes were grown to near confluence and then incubated (d0) in serum-free differentiation medium [2 µmol/l rosiglitazone, 25 nmol/l dexamethasone, 0.5 mmol/l methylisobuthylxantine, 0.1 µmol/l cortisol, 0.01 mg/ml transferrin, 0.2 nmol/l triiodothyronine, and 20 nmol/l human insulin]. After 4 days, the medium was changed, and the cells were further cultured in medium supplemented with 0.1 µmol/l cortisol, 0.01 mg/ml transferrin, 0.2 nmol/l triiodothyronine, and 20 nmol/l human insulin. TBT-exposed SGBS preadipocytes were differentiated as specified above; for the initial four days (d0–d4), the differentiation medium was additionally supplemented with TBT (25 nM). Cells were harvested at d10 of differentiation.

Transcriptomic analysis was performed using Agilent microarrays to determine differentially expressed genes (DEGs) between treatment groups. Samples collected for metabolomics were analysed using Agilent 1290 Infinity HPLC LC System coupled to an Agilent 6540 HRMS-QTOF/ LCMS system with Reversed-Phase (RP) and Hydrophilic Interaction (HILIC) Liquid Chromatography columns in positive (PI) and negative ionisation (NI) modes. Data preprocessing, batch correction, and statistical analyses were conducted in R using limma for transcriptomics and xcms, IPO, PMCMRplus, and xMSannotator packages for metabolomics for univariate analyses while for multivariate analyses utilised the mixOmics package. Joint pathway analyses were carried out by importing data from univariate and multivariate analyses, thus leveraging the capabilities of the MetaboAnalystR package to facilitate a better comprehension of multi-omic mechanisms.

The univariate statistical analysis resulted in 1088 differentially expressed genes (DEGs) between TBT-exposed cells compared to controls while 551 and 239 metabolites were statistically significantly differentially detected from HILIC and RP analysis. Overall, the categories of glycerophospholipids and fatty acyls were most commonly detected. Across all analyses, the top three commonly occurring perturbed pathways were glycerophospholipid metabolism, retrograde endocannabinoid signalling, and taurine and hypotaurine metabolism. Glycerophospholipids, a class of lipids fundamental to cellular membranes and signalling, play a pivotal role in various physiological processes, including energy homeostasis, lipid transport, and inflammation, and serve as precursors for signalling molecules that directly modulate metabolic processes like inflammation and insulin sensitivity (Alves et al., 2021; Yoon et. al., 2021). Furthermore, aberrant glycerophospholipid metabolism is implicated in the development of non-alcoholic fatty liver disease (NAFLD), a condition closely linked to metabolic disturbances (Liang & Dai, 2022). Disruption in retrograde endocannabinoid signalling has recently been associated with dyslipidemia, obesity and diabetes by influencing cannabinoid receptors both centrally and peripherally (Lipina et al., 2023). Moreover, the endocannabinoid system in general is suggested to play a crucial role in oxidative/nitrosative stress and inflammatory processes within the liver, and thus, been linked to NAFLD and metabolic dysfunction (Jorgaĉević et al., 2021). Taurine and hypotaurine play vital roles in maintaining normal lipid metabolism and is also involved in antioxidant and bile acid conjugation activities (Du et al., 2020). In addition, hepatic taurine levels could potentially influence hepatic steroid metabolism and other lipid metabolisms (Du et al., 2020). Another study by Miao et al., (2022) found that taurine and hypotaurine supplementation was shown to significantly alter outcomes of liver injury, fat accumulation, and insulin sensitivity in NAFLD mice.

Overall, the integrated omics analysis pipeline demonstrated the complexity of cellular responses to TBT exposure, utilizing both univariate and multivariate analytical methods to deepen our understanding of metabolic mechanisms. This approach not only highlighted individual molecular changes but also revealed their collective impact on cellular pathways. Key findings include the consistent identification of perturbed pathways such as glycerophospholipid metabolism, retrograde endocannabinoid signaling, and taurine and hypotaurine metabolism. These insights lay the groundwork for future mechanistic studies and could guide the development of strategies to mitigate the adverse effects associated with exposures to obesogenic and endocrine disrupting compounds with regards to MetS.

References

ALVES, M. A., LAMICHHANE, S., DICKENS, A., MCGLINCHEY, A., RIBEIRO, H. C., SEN, P., WEI, F., HYÖTYLÄINEN, T. & OREŠIČ, M. 2021. SYSTEMS BIOLOGY APPROACHES TO STUDY LIPIDOMES IN HEALTH AND DISEASE. BIOCHIMICA ET BIOPHYSICA ACTA (BBA) - MOLECULAR AND CELL BIOLOGY OF LIPIDS, 1866, 158857.

DU, J. E., YOU, Y. A., KWON, E. J., KIM, S. M., LEE, J., HAN, K. H., & KIM, Y. J. (2020). MATERNAL MALNUTRITION AFFECTS HEPATIC METABOLISM THROUGH DECREASED HEPATIC TAURINE LEVELS AND CHANGES IN HNF4A METHYLATION. INTERNATIONAL JOURNAL OF MOLECULAR SCIENCES, 21(23), 9060.

JORGAĈEVIĆ, B., VUĈEVIĆ, D., SAMARDŽIĆ, J., MLADENOVIĆ, D., VESKOVIĆ, M., VUKIĆEVIĆ, D., ... & RADOSAVLJEVIĆ, T. (2021). THE EFFECT OF CB1 ANTAGONISM ON HEPATIC OXIDATIVE/NITROSATIVE STRESS AND INFLAMMATION IN NONALCOHOLIC FATTY LIVER DISEASE. CURRENT MEDICINAL CHEMISTRY, 28(1), 169-180.

LIANG, K. & DAI, J. Y. 2022. PROGRESS OF POTENTIAL DRUGS TARGETED IN LIPID METABOLISM RESEARCH. FRONT PHARMACOL, 13, 1067652.

LIPINA, C., RASTEDT, W., IRVING, A. J., & HUNDAL, H. S. (2012). NEW VISTAS FOR

TREATMENT OF OBESITY AND DIABETES? ENDOCANNABINOID SIGNALLING AND METABOLISM IN THE MODULATION OF ENERGY BALANCE. BIOESSAYS, 34(8), 681-691. HTTPS://DOI.ORG/10.1002/BIES.201200031.

MIAO, J., GUO, L., CUI, H., WANG, L., ZHU, B., LEI, J., ... & ZHANG, Z. (2022). ER-CHEN DECOCTION ALLEVIATES HIGH-FAT DIET-INDUCED NONALCOHOLIC FATTY LIVER DISEASE IN RATS THROUGH REMODELING GUT MICROBIOTA AND REGULATING THE SERUM METABOLISM. EVIDENCE-BASED COMPLEMENTARY AND ALTERNATIVE MEDICINE, 2022

YOON, H., SHAW, J. L., HAIGIS, M. C. & GREKA, A. 2021. LIPID METABOLISM IN SICKNESS AND IN HEALTH: EMERGING REGULATORS OF LIPOTOXICITY. MOL CELL, 81, 3708-3730.

169bf Modeling Ionic and Sequence Effects on the Swelling Behavior of Polyampholyte Brushes

Debadutta Prusty, Alejandro Gallegos, Jianzhong Wu

The structural responses of polyampholytes to various chemical stimuli are sensitive to the sequence of monomeric groups that bear positive and negative charges. However, a theoretical understanding is yet to be established to account for the sequence effects under diverse environmental conditions such as solution pH, salt concentration, and ion valence. As a first step towards delineating the physics, we consider strong polyampholyte brushes with either a diblock or an alternating chain architecture and study their structure and swelling behavior in response to variations in salt concentration, ion size and valency using a polymer density functional theory. As the salt concentration increases, an alternating brush displays a re-entrance behavior, characterized by an initial reduction of the brush height followed by a subsequent brush expansion. The impact of ion size becomes appreciable only at high salt concentrations, wherein smaller ions contribute to increased brush heights. For a diblock brush, however, the brush height decreases monotonically with the addition of salt when the counterions to the lower block are significantly larger than the polymer segments. A re-entrance behavior is observed only when the ion diameter is equal to or smaller than that of the monomers. The non-monotonic trend contradicts the so-called anti-polyelectrolyte effect, which predicts an expansion of the polyampholyte brush upon the addition of salt. While the variation of the counterion size for the upper block generates a similar swelling behavior, the ion size has opposite effects on the degree of inter-block association. These trends have been rationalized by considering the interplay of excluded-volume interactions with electrostatic and correlation effects. For both types of polyampholyte brushes, the salt effect becomes significantly more pronounced in the presence of divalent ions, with the alternating brush displaying a stronger dependence on the salt concentration.

169bg Can Classical Nucleation Theory Describe Heterogeneous Crystal Nucleation on Non-Uniform Surfaces?

Fernanda Vargas, Sarwar Hussain, Amir Haji-Akbari

Crystal nucleation stands as a pivotal phenomenon in numerous natural and industrial processes. Its profound implications extend to high-end industries, including semiconductors, solar cells, and aviation, where the understanding alongside the manipulation of heterogeneous crystal nucleation is of utmost importance as it is at the heart of the production of single crystalline materials. The time and length scales at which nucleation proceeds are too small to capture using most experimental techniques, presenting a challenge to experimental investigations of nucleation. As such, computer simulations are utilized to unveil the molecular details of the nucleation process. Historically, classical nucleation theory (CNT) has been used extensively to describe the physics of both homogeneous nucleation and heterogeneous nucleation. Despite its success, its validity or lack thereof under different circumstances is still debated.

In this work, we use molecular simulations and advanced path sampling techniques to probe heterogeneous nucleation of the Lennard-Jones (LJ) crystal on different uniform and patterned surfaces. To rigorously assess the applicability of CNT, we first compute heterogeneous nucleation rates using the jumpy forward flux sampling algorithm and examine their dependence on temperature. We also assess the stability of the FCC and HCP crystals next to each surface by conducting conventional MD simulations of FCC and HCP crystals at its vicinity and characterizing the extent of pre-melting. Finally, we also analyze crystalline nuclei along the nucleation pathway and find them to change shape upon enlargement, directly contradicting one of the core assumptions of CNT that crystalline nuclei are spherical caps with fixed contact angles. Overall, our findings underscore the nuances and intricacies of applying CNT to probe the kinetics of heterogeneous nucleation.

169bh Computational Investigation of Reaction Coordinate Optimality for Ice Nucleation Studies

Kimia Sinaeian, Amir Haji-Akbari

Crystallization plays a crucial role in a myriad of processes spanning from cloud microphysics to the production of pharmaceuticals and semiconductors, underscoring its profound significance in both microscopic and macroscopic processes. As a first-order phase transition, crystallization typically proceeds through the nucleation and growth mechanism. Within this framework, nucleation is the initial step in crystallization. As such, understanding nucleation at a microscopic level, is of intense interest for advancing our ability to understand and engineer all such processes. More often than not, nucleation is the rate-limiting step in crystallization.. This makes nucleation a rare event that occurs over short timescales but only after considerably longer waiting times. Conventional molecular simulation techniques fail to sample the right configurations that successfully trespass large energy barriers characteristic of the nucleation process. As such, advanced sampling techniques are needed for efficient exploration of the nucleation pathway. One necessary ingredient of such techniques is the availability of suitable collective variables (CVs)- also known as order parameters- that accurately describe the progress of nucleation.

In this work, we use molecular dynamics simulations with forward flux sampling (FFS), and machine learning to assess conventional CVs for homogeneous and heterogeneous ice nucleation. By analyzing a collection of configurations with diverse committer probability values, we screen tens of thousands of distinct CVs that construct crystalline nuclei using different criteria that employ five cut-offs and/or decision points. In addition to standard cut-offs used for constructing bond order parameters (BOPs), we assess the efficacy of two other strategies, namely chain exclusion and the inclusion or exclusion of the first hydration shell around a crystalline nucleus. We find these techniques to not culminate in more potent CVs, although their efficacy or lack thereof depends on the key cut-offs used for constructing BOPs.

169bi Hydration Free Energies of Linear Alkanes (C1-C20) and Surfactants in Different Water Models

Yalda Ramezani, Sumit Sharma

We report differences in the hydration free energies of alkanes ranging from methane to eicosane (C20) between different water models (SPC/E, TIP4P2005, OPC) calculated using the free energy perturbation method in molecular simulations. We find that the tail correction for the potential function and the potential cut-off distance associated with the alkane-water interactions have significant impact on the calculated hydration free energies. We also calculate the hydration free energy of cationic and charge-neutral surfactants with differently sized polar head groups. We find that cationic surfactants have a much higher affinity to get solvated in water as compared to charge-neural surfactants. The cationic surfactants are associated with formation of a water finger at the water-air interface, which changes the shape of the free energy profile near the interface. Lastly, we show how the solubility of surfactants can be significantly enhanced by changing the single substitutions on the alkyl tail. These results reveal the contributions of the polar head and the alkyl tail on the overall hydration free energy, and thus are useful for designing surfactants with the desired aqueous solubility needed for various applications.

169bj Size Is an Important Factor in Partitioning of Cargo Molecules into Liquid Condensates and Interfaces

Arjun Singh, Fleurie Kelley, Benjamin S. Schuster, Gregory Dignon

Within biology, compartmentalization can be achieved through lipid membranes serving as barriers, or through the process of liquid phase separation of biomolecules. Compartments arising through the process of phase separation are generally termed biomolecular condensates (BMCs) and are formed by the attractive interactions between various biomolecules, particularly proteins and nucleic acids. Recent work has demonstrated that BMCs are permeable to many biomolecules, and that various factors may promote or oppose incorporation into a condensed proteinaceous droplet. We find that larger cargo molecules are more excluded from BMCs, and that even weak adhesive forces can drive them to associate with the surface. Using simulations and theory, we explore the interplay between attractive forces and surface enrichment inside condensates toward a general framework of predicting cargo partitioning into BMC interiors and interfaces, in agreement with experimental measurements.

169bk Utilizing Surfactant-Specific Graph Convolutional Networks to Predict Surfactant Adsorption Efficiency

Seokgyun Ham, Xin Wang, Rui Qiao

Quantifying the adsorption efficiency of surfactants, measured by the surface area occupied per molecule, is vital for the development of new surfactants, particularly those for firefighting foams. Dense surfactant packing at interfaces is believed to significantly hinder oil movement through the surfactant layer due to high resistance. Nevertheless, deciphering the intricate link between a surfactant's structural makeup and its adsorption efficiency poses a significant challenge. To address this challenge, we introduce a graph convolution network (GCN) specifically designed for surfactants to predict their adsorption efficiency. This model categorizes the surfactant molecules into head and tail sections, incorporating both specific features of these sections and general atomic characteristics into the feature matrix. Some of these head and tail features are derived from low-cost molecular dynamics (MD) simulations. Furthermore, the model is trained on a comprehensive dataset of 98 different surfactant structures, covering all surfactant categories such as anionic, cationic, zwitterionic, and nonionic, thus ensuring its applicability across various surfactant types. Our findings indicate that this specialized model surpasses conventional deep learning models in accuracy when predicting surfactant adsorption efficiency across a wide array of data. Additionally, molecular saliency maps are explored to gain insights into how specific surfactant substructures affect adsorption efficiency.

169bl Maxwell-Stefan Diffusivities of Oil-CO_2 Mixtures in Nanopores: Physics and Machine Learning Models

Hongwei Zhang, Xin Wang, Rui Qiao, Qinjun Kang, Bicheng Yan, Shuyu Sun

Transport properties of oil and CO_2 confined in nanoscale pores are key inputs for physics-based modeling of CO_2-enhanced oil recovery (EOR) in unconventional reservoirs. Due to confinement and fluid-wall interactions, oil and CO_2 molecules exist as heterogeneous mixtures across nanopores. Their transport properties can deviate from those of bulk mixtures, but the corresponding models for nanoconfined liquid mixtures are lacking. This study investigates the Maxwell-Stefan (MS) diffusivities of CO_2-C10 (1: CO_2; 2: C10) mixtures in nanopore under the compositions relevant to CO_2 Huff-n-Puff by molecular dynamics (MD) simulations.

In the composition space explored here, D_{12} (characterizing the CO_2-C10 interaction) is insensitive to the mixture composition in contrast to that without nanoconfinement. $D_{1,s}$ (CO_2-wall interaction) increases sharply with CO_2 loading, while a nonmonotonic dependence on C10 loading is observed on the $D_{2,s}$ (C10-wall interaction). Also, surprisingly, opposite to the expectation for dense fluid mixtures confined in nanopores, $D_{2,s}$ is negative. These observations can ultimately be traced to the fact that CO_2 molecules have far stronger wall affinity than C10 molecules to the calcite wall, resulting in significantly heterogeneous density distribution and ultra-low mobility of the adsorbed CO_2 molecules.

As MD simulations are computationally expensive, a Multi-Output Gaussian Process machine learning model is developed as a surrogate model to predict the MS diffusivities in the vast composition space efficiently. Trained from a limited MD dataset, the model achieves a less than 10% relative root mean square error in the specified composition space.

169bm Protein Mediated Calcite Nucleation and Growth Characterized with Molecular Dynamics Simulations

Marlo Zorman, Harley Pyles, Fatima Davila, James J. De Yoreo, David Baker, Christopher J. Mundy

Biomineralization is a process by which organisms produce complex and multifunctional inorganic materials via directed crystal growth. This process displays a level of hierarchical order and control beyond what is capable with contemporary materials design and therefore presents an exciting opportunity for study. Materials formed through biomineralization are ubiquitous in nature; one of the most common and well-studied examples is the protein-mediated formation of carbonate shells by marine animals. Advancements in protein design have resulted in the discovery of proteins that can imitate complex processes such as carbonate shell formation. A more complete understanding of the molecular-level interactions between proteins and surfaces that regulate these biomineralization processes would provide insight necessary for better predictive design. One model that could provide such insight is the protein mediated growth of calcium carbonate, and in particular the growth of the most stable polymorph of calcium carbonate: calcite. Physics based simulations offer a particularly attractive opportunity to characterize these mechanisms at the atomic scale. In this work, we use classical Molecular Dynamics and Monte Carlo simulations investigate the mechanism by which two de novo designed proteins, FD31 and DHR49, mediate and direct calcite nucleation and growth. The results of these simulations provide insight into 1) the specific residues that promote nucleation and binding, and 2) the role of water in mediating these interactions. Finally, we hypothesize a mechanism by which these proteins direct calcite growth. This work provides insight into the mechanisms of biomineralization and presents potential avenues for improvement of de novo protein design.

169bo Protein Preparation: Is One Protonation State Enough?

Prerna Gera, Nate Abraham, Mossa Ghattas, Vibhas Aravamuthan, Sarah Williams, Tom Kurtzman, Jason Burch, Dan McKay

Preparation of experimental protein crystal structures for molecular simulations is a non-trivial task when attempting to recapitulate a system. Manual preparation of a protein requires the user to identify all possible rotamer and protonation unknowns and evaluate the likelihood of each state to come up with a possible solution[1]. Using unrestrained MD, we would validate our model or, more often than not, iterate on the model to improve the match with X-ray. For certain proteins this process could be completed within a day, but for more intricate systems it could take a few weeks. Alternatively, automated methods, such as SPRUCE[2], expedite the process down to a few minutes to generate a singular model to validate with unrestrained MD. Regardless, all the protein preparation work described thus far is predicated on the assumption that there is a single state of the protein despite the complexity of the environment in which the protein exists.

By encompassing our expert knowledge from manual protein preparation, we've built a tool to enumerate multiple protonation and rotameric states efficiently. Even if we were to assume a single protein state is sufficient, we currently do not have access to all feasible states in which to identify "the best" singular one. With several amino acids having multiple protonation and/or rotameric state, it can be quite daunting to see this problem as a combinatorial explosion. By exploiting the local environment we can evaluate the hydrogen bonding energy around an ambiguous residue to identify unlikely states and reduce the number of possible outcomes. At the end, there may be a handful of residues with multiple solutions that we use to generate all possible states, all of which can be completed within minutes. Using unrestrained molecular dynamics we then can get an understanding of how the varied protonation and rotameric states recapitulate the X-ray structure.

Multiple rotamer and protonation states can exist that are consistent with the X-ray structure, therefore making the selection of a single state arbitrary. We evaluate the rotamer and protonation states of 68 proteins from the DUD-E set[3] and refine the structures using 10ns of unrestrained molecular dynamics to understand the number of possible states and how they relate back to the X-ray. For most protein systems there are less than 10 possible states found in the preparation, with the maximum just over 250 (black squares in **Figure 1**). Although there are many potential states for some proteins, they are still trackable, nullifying the assumption of this being a combinatorial explosion. Furthermore, the average structure from each simulation was compared to the X-ray structure and the RMSD was calculated for 50, 70, and 90% of the atoms by B-factor with the number of states less than 1.0Å reported (**Figure 1**). In most of the proteins there are multiple states where 50-70% of the atoms by B-factor match the X-ray with an RMSD < 1.0Å, underscoring how the model varies widely and selection of a single state is arbitrary.

These findings have downstream consequences to understanding the protein and the selection of a single state can be detrimental when attempting to discover novel chemical matter. Given a fixed protein conformation from the X-ray in this study, we can see how residues proximal to the binding pocket vary and can influence the environment quite dramatically. In several PDB entries we found that there are residues adjacent to significant pockets that have multiple stable states throughout the 10ns unrestrained MD simulation. In all cases the varying states lead to dramatic changes in local environment with acceptor, donors, or lipophilic groups all being stable. Using a singular state would result in critical information missing when making design choices, undermining the predictive nature of the methods.

This work is just the start of a much larger conversation around protein preparation but provides a much more promising outlook on enumerating protein states. In all cases explored here there are multiple rotameric and protonation states that vary and still recapitulate the X-ray structure, but all are still tractable given the sufficient compute resources. For the purpose of optimization or exploring a fixed protein conformation, one could easily study all relevant states and make more informed decisions during the design process. Expanding this concept, we can see how these different states can impact sampling when it comes to protein motion or the conformational landscape when running long MD.

Figure 1: For 68 protein systems selected form the DUD-E set[3] this shows the total number of states determined by our method (black squares), the total number of these states which produce a time averaged structure after 10ns of unrestrained MD with an RMSD of less than 1 for lowest 50% of the atom in the original X-ray structure selected by B-factor (red diamonds), the total number of states with RMSD less than 1 for the lowest 70% by B-factor (blue triangles) and the total number of states with RMSD<1 for the lowest 90% by B-factor (brown circles).

References

1: Bender, B.J., Gahbauer, S., Luttens, A. *et al.* A practical guide to large-scale docking. *Nat Protoc* **16**, 4799–4832 (2021). https://doi.org/10.1038/s41596-021-00597-z

2: SPRUCE: OpenEye tools

3: DUD-E Mysinger MM, Carchia M, Irwin JJ, Shoichet BK J. Med. Chem., 2012, Jul 5. doi 10.1021/jm300687e .

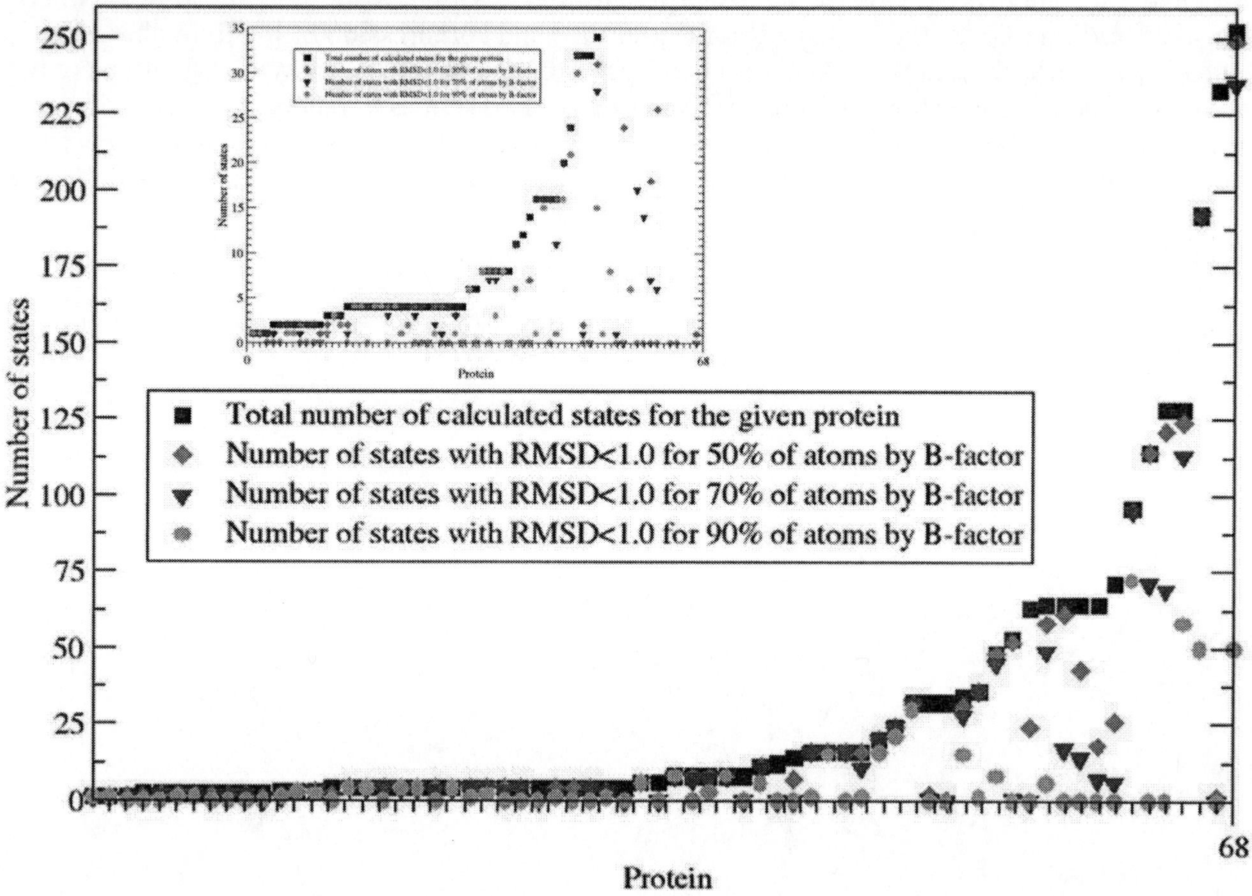

169bp Theoretical Approach to Elucidating the Dynamics Behavior of PFAS at the Water and Hydrophobic Deep Eutectic Solvents Interphase

Mert Atilhan, Noor Alomari

hydrophobic deep eutectic solvents (HDES) were investigating their potential for per- and polyfluoroalkyl substances (PFAS) remediation from wastewater through a nanoscopic approach using quantum chemistry and classical molecular dynamics simulations. we study the properties of two selected widely used long-chain PFAS, Perfluorooctanoic acid (PFOA) and perfluorooctanesulfonic acid (PFOS), solutions in H2O with a novel HDES obtained by mixing cineole (CIN) and linoleic acid (LNA). The result showed the interaction between PFAS and an innovative HDES at equimolar ratios and the interphase diffusion of PFAS through the HDES is a multi-stage process involving diffusion, interface sorption, and subsequent migration into the bulk aqueous phase. The simulations underscore a robust affinity between the PFAS molecules and the HDES, with the displacement of PFAS from the water interface marked by the formation of complex hydrogen-bonded networks. This intricate interplay results in a significant reorganization of water molecules, leading to potential clustering at the interface and compression of the fluid phase.The liquid structures are characterized by strong hydrogen bonding in the PFAS hydroxyl groups with the O atoms in CIN and the carbonyl groups in LNA. Our study provides a molecular-level understanding of the PFAS extraction mechanism and suggests that the use of HDES solvents with low toxicity as extraction agents for PFAS remediation from wastewater could be a proper solution to design novel PFAS treatment materials suitable for liquid extraction.

169bq A Nanoscopic Explanation on Hydrophobic Deep Eutectic Solvents and Their Carbon Dioxide Solubility Performance at High-Pressure.

Noor Alomari, Mert Atilhan

This study investigates the development, characterization, and application of natural hydrophobic deep eutectic solvents (HDES) in high-pressure gas capture, focusing specifically on carbon dioxide (CO_2) and nitrogen (N_2). Two binary HDES systems were formulated, using LnA as a hydrogen bond donor (HBD) and CAR and CIN as hydrogen bond acceptors (HBA). High-pressure gas solubility through integrated computational approach was conducted to assess CO_2 and N_2 sorption capacities. COSMO-RS was used to predict solid–liquid equilibria (SLE) and the eutectic point of the NADES, while Density Functional Theory (DFT) calculations aided in optimizing molecular geometries and understanding the dispersion interactions. Furthermore, molecular dynamics (MD) simulations were employed to evaluate solubility of CO_2 and N_2 in the NADES and their interaction energies. The results from this computational approach show the potential of these HDES systems in high-pressure gas capture.

169br Effect of Maturity on the NMR Relaxation of Kerogen Using MD Simulations

Yunke Liu, Arjun Valiya Parambathu, George J. Hirasaki, Walter Chapman, Philip Singer

Abstract:

Maturation studies of organic-rich source rocks are integral to evaluating unconventional hydrocarbon reservoir formations, including gas shale and oil shale. Kerogen, the predominant organic macromolecular polymer within these source rocks, holds significant importance in unconventional formation evaluation. Kerogen-rich gas shale reservoirs play a pivotal role during the energy transition, serving as a clean-burning energy source and a possible sink for geological carbon capture and storage (CCUS).

We previously showed how atomistic molecular dynamic (MD) simulations can elucidate the underlying physics behind the ^{1}H NMR relaxation of bulk fluids [1-4] and fluids under organic nanoconfinement [5-6], without resorting to the traditional assumptions of BPP theory. More specifically, we used MD simulations to compute the ^{1}H-^{1}H dipole-dipole autocorrelation function, which successfully predicted the ^{1}H NMR relaxation dispersion (i.e., frequency dependence) of fluids, without any relaxation models and without any adjustable parameters in the interpretation of the simulations.

In this work we extend our MD simulation technique to study the transverse ^{1}H NMR relaxation (T_{2G}) of the solid organic matrix itself, i.e., kerogen, with varying thermal maturity. The MD simulations are used to quantify both the intramolecular and intermolecular components of T_{2G} relaxation by computing the ^{1}H-^{1}H dipole-dipole second moment (SM) of the kerogen molecules with varying thermal maturity. We find a strong correlation between SM and H/C (elemental ratio) of kerogen as a function of maturity, which we use to predict T_{2G} (and T_1) relaxation versus H/C. We find a consistent trend between the simulated T_{2G} (and T_1) versus H/C, and the trend found from NMR measurements of Type II-S organic rich chalk as a function of maturity using a solid-echo pulse sequence to detect the solid kerogen [7].

Illustration of the uploaded figures:

(a) Schematic of the simulation details for Type IIC Kerogen as an example. 8 Type IIC Kerogen molecules are packed to conduct the MD simulation.

(b),(c) MD simulation of the intra-molecular $G_R(t)$ (left) and inter-molecular $G_T(t)$ auto-correlation functions for Type IA, IIA, IIB, IIC, IID, and IIIA Kerogen.

(d) Computed the intra and inter square root of second moment in kHz of Type IA, IIA, IIB, IIC, IID, and IIIA Kerogen as a function of H/C using the autocorrelation function through the MD simulation. As a comparison, the results for benzene (aromatic) and heptane (aliphatic) are also shown.

(e) The predicted longitudinal relaxation rate (R1 = 1/T1) by MD simulation and Plateau model (red) and compared the experimental R1 of Type IIS organic-rich chalk (blue) as a function of H/C. The consistent trend is observed between the simulated T2G (and T1) versus H/C, and the trend found from NMR measurements of Type II-S organic rich chalk as a function of maturity using a solid-echo pulse sequence to detect the solid kerogen.

References:

1. Singer et.al, *J. Magn. Reson.* **2017**, *277*, 15–24, DOI: 10.1016/j.jmr.2017.02.001
2. Singer et.al., *J. Chem. Phys.* **2018**, *148*, 164507, DOI: 10.1063/1.5023240
3. Singer et.al, *J. Chem. Phys.* **2018**, *148*, 204504, DOI: 10.1063/1.5027097
4. Asthagiri, et.al, *J. Phys. Chem. B* **2020**, *124*, 10802–10810, DOI: 10.1021/acs.jpcb.0c08078
5. Parambathu et.al, *J. Phys. Chem. B* **2020**, *124*, 3801–3810, DOI: 10.1021/acs.jpcb.0c00711

6. Parambathu et.al, *J. Phys. Chem. Lett.* **2023**, *14, 4,* 1059–1065, DOI: 10.1021/acs.jpclett.2c03699
7. Liu et.al, *Fuel* **2024**, *367*, 131378, DOI: 10.1016/j.fuel.2024.131378

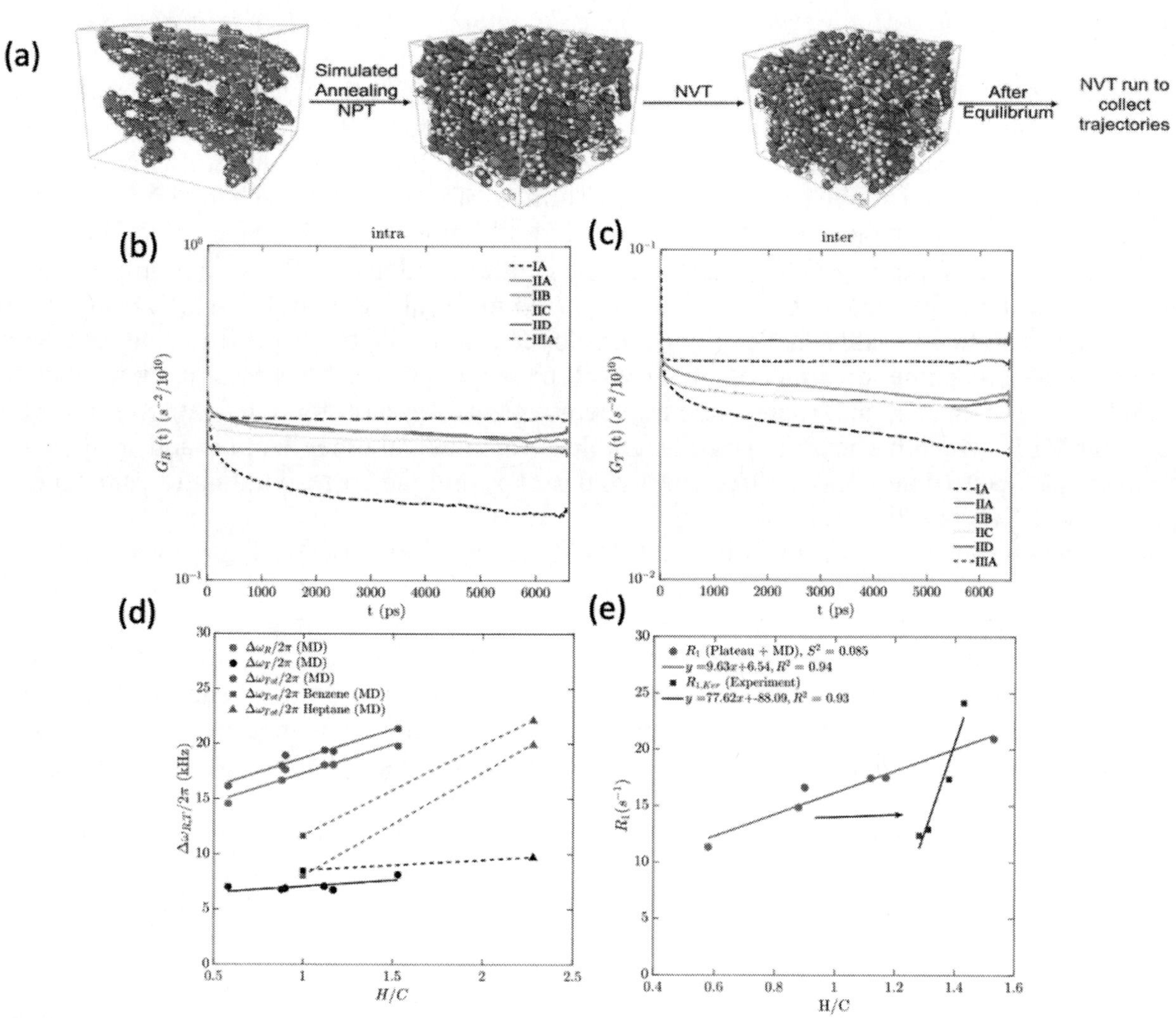

169bs Elucidating the Fluxionality and Dynamics of Zeolite-Confined Au Nanoclusters Using Machine Learning Potentials

Siddharth Sonti, Chenghan Sun, Zekun Chen, Davide Donadio, Surl-Hee Ahn, Ambarish Kulkarni

Sub-nanometric clusters (NCs) of transition metal (TM) atoms, typically consisting of fewer than 15 atoms, have exhibited remarkable catalytic activity in various industrial reactions. However, these NCs are thermodynamically unstable and susceptible to deactivation due to sintering. Previous experiments have proposed zeolites as effective structural scaffolds to stabilize small NCs. However, key physiochemical phenomena such as fluxionality and diffusion properties of such zeolite-confined TM NCs (TM@zeolites) are not well understood. The central challenge is the steep computational cost associated with performing sufficiently long *ab initio* molecular dynamics (AIMD) simulations. As a step towards addressing this challenge, we developed a self-adaptive workflow that leverages two state-of-the-art machine learning potential (MLP) packages (i.e., SchNetPack and neuroevolution potential) to develop an accurate, robust, and transferable MLP for a prototypical Au@zeolite system. The resulting MLP, which is shown to be transferable across several other zeolites and various temperatures, is used to determine the free energy landscape and the corresponding rates for the diffusion of gold NCs in LTA zeolites.

Sub-nanometric clusters (NCs) of transition metal (TM) atoms, typically consisting of fewer than 15 atoms, have exhibited remarkable catalytic activity in various industrial reactions. However, these NCs are thermodynamically unstable and susceptible to deactivation due to sintering. Previous experiments have proposed zeolites as effective structural scaffolds to stabilize small NCs. However, key physiochemical phenomena such as fluxionality and diffusion properties of such zeolite-confined TM NCs (TM@zeolites) are not well understood. The central challenge is the steep computational cost associated with performing sufficiently long *ab initio* molecular dynamics (AIMD) simulations. As a step towards addressing this challenge, we developed a self-adaptive workflow that leverages two state-of-the-art machine learning potential (MLP) packages (i.e., SchNetPack and neuroevolution potential) to develop an accurate, robust, and transferable MLP for a prototypical Au@zeolite system. The resulting MLP, which is shown to be transferable across several other zeolites and various temperatures, is used to determine the free energy landscape and the corresponding rates for the diffusion of gold NCs in LTA zeolites.

169bt Combining Forward and Inverse Design of Covalent-Organic Frameworks for Methane Storage Via Data-Driven Discovery

Xiaoyu Wu, Jianwen Jiang

This study introduces a new strategy for the forward and inverse design of Covalent-Organic Frameworks (COFs) aimed for optimal methane (CH_4) storage. By computationally synthesizing CH_4 deliverable capacities in COF databases, our approach distinguishes itself through a dual featurization model that captures both framework and pore chemistries. Sparse chemical features, such as the presence of specific bond types, are adeptly managed through factorization machines, while continuous variables like density and pore sizes benefit from deep learning methodologies. This fusion significantly advances prediction accuracy for COFs with high CH_4 storage potential.

Inverse design poses a unique challenge for COFs, given their varied connection types and similar chemical identities of secondary building blocks (SBUs). To address this, we employed a Crystal Diffusion Variational Autoencoder (CDVAE), adopting a Structure is All You Need philosophy. This approach ensures that every component, including SBUs and topologies, remains flexible while navigating the chemical space, relying entirely on the encoded space. With a focused training regime on top-performing COFs guided by property optimizations, we aim to facilitate the efficient design of high-performing COFs for CH_4 storage.

Our work transcends traditional boundaries in COF research, offering synergistic forward and inverse design to fast-track the discovery of new COF for high CH_4 storage. We believe this work shall lay a solid groundwork for future advancement in gas storage.

169bu Hydrophobic Deep Eutectic Solvents As Extraction Agents of Nitrophenolic Removel from Water

Noor Alomari

In this study, we focus on hydrophobic deep eutectic solvents (HDES) and investigate their potential for Nitrophenol-based chemicals. These pollutants are used in agriculture as fungicides and pesticides and contribute to groundwater contamination. These compounds are also commonly used for making dyes, rubber mate- rials, and paint coloring. Nitrophenol compounds are known to cause problems with respiratory, ocular, and hematological-related health issues. Due to their environmental and health impacts, the World Health Organization has listed these compounds as one of the most widely observed pollutants that must be addressed because of that negatively impact human health and the environment. density functional theory (DFT) and classical molecular dynamics (MD) simulations were carried out to study close-range interactions and how HDES and phenolic contaminants interact with each other in the bulk phase. Radial and spatial distribution functions, hydrogen bonding distribution, and energy profiles were studied in detail. The results show high extraction efficiency of up to 95% at short HDES exposure time to contaminated water with both phenolic compounds considered in this work.in addition to computational work,experimental of three different types of monoterpenoid and fatty acid-based HDES were used in liquid–liquid extraction to remove varying concentrations of 2- and 4-nitrophenol compounds from water. Liquid-liquid extractions were performed at various temperatures and mixing times to optimize the extraction performance. Ultraviolet– visible (UV–Vis) spectrophotometry and mass spectroscopy were utilized to determine and confirm the exaction efficiencies of the various HDES systems.

169bx Direct Simulation of Supported Ag Nanoparticles Via Machine Learning Interatomic Potentials

Tristan Maxson, Tibor Szilvasi

Supported nanoparticles have a wide range of applications (thermal-, electro-, and photocatalysis, sensors, etc.) but their in-situ characterization and computational modeling prove challenging. Current nanoparticle models typically use low-index periodic models with defect-free surface due to their lower computational cost compared with nanoparticle models. The recent development of machine learning interatomic potentials (MLIPs) can provide computationally viable approximations of density functional theory calculations while maintaining similar levels of accuracy in energies and forces. We show our nanoparticle models are both accurate and computationally viable by using equivariant MLIPs. To benchmark the validity of MLIPs for nanoparticle simulations, we build upon the two recent works of Campbell et al[1,2] that give experimental understanding of the adhesion of Ag nanoparticles to Ni(111) supported graphene and their chemical potential as a function of particle size. In this work, we train MLIPs via an iterative committee training scheme to identify poorly predicted structures and improve the training set. Simulated Annealing (SA) and hybrid Monte Carlo Molecular Dynamics (MCMD) simulations are performed to explore chemically relevant geometries of nanoparticles at various temperatures to make comparison to Campbell's results. We demonstrate structural deviations from the idealized Winterbottom construction due to the finite size effects at particle sizes below approximately Ag_{800}. The nanoparticles are then analyzed to provide insights into the distributions of structures that may be present. This work marks progress toward understanding the structural properties of supported nanoparticles under experimentally relevant conditions as we move from the nanocluster (diameter below 1 nm) to the nanoparticle size regime (diameter above 1 nm).

(1) Rumptz, J. R.; Mao, Z.; Campbell, C. T. Size-Dependent Adsorption and Adhesion Energetics of Ag Nanoparticles on Graphene Films on Ni(111) by Calorimetry. *ACS Catalysis*. **2022**, *12*, (5), 2888-2897.

(2) Zhao, K.; Auerbach, D. J.; Campbell, C. T., Calorimetric Energies and Chemical Potentials of Metal Atoms in Catalytic Nanoparticles on Oxide and Carbon Supports: Improved Size Dependencies and Adhesion Energies. *ACS Catalysis*, **2023**, *13*, (21), 13968-13981.

169bz Accelerating Polymer Informatics Via Polymer Similarity

Jiale Shi, Debra J. Audus, Bradley Olsen

Developing accurate similarity algorithms among chemical entities is an essential task in polymer informatics, facilitating necessary functions such as regression to predict polymer properties, classification to identify promising candidates and ranking of polymer database search results. Unlike small molecules with well-defined molecular structures and biomacromolecules with specific sequences, synthetic polymers are more complex, typically characterized by ensembles of polymer chains that exhibit variations in length, sequence, composition, topology, and quantity. Since the existing similarity methods used for small molecules and sequence-defined biomacromolecules require well-defined deterministic structures, they cannot be directly utilized for polymer similarity calculations. Therefore, despite its importance, pairwise chemical similarity for polymers remains an open problem.

Even if the exact distributions are not known, the similarity for a synthetic polymer can still be computed. First, the synthetic polymer is represented by BigSMILES, a structurally-based line notation for describing macromolecules. Then, this BigSMILES string is canonicalized and converted to a graph-based representation. Next, the stochastic graph representation is separated into three parts: repeat units, end groups, and polymer topology. The earth mover's distance is utilized to calculate the similarity of the repeat units and end groups, while the graph edit distance is used to calculate the similarity of the topology. These three values can be linearly or nonlinearly combined to yield an overall pairwise chemical similarity score for polymers that is largely consistent with the chemical intuition of expert users and is adjustable based on the relative importance of different chemical features for a given similarity problem. [Shi et al. Macromolecules 2023, 56, 18, 7344-7357]

When polymers are represented as an ensemble where its distributions are characterized, earth mover's distance metric is proposed to calculate the pairwise similarity score between two polymer ensembles. The power of using earth mover's distance to characterize the pairwise similarity score between polymer ensembles is illustrated in four examples, including two-chain copolymer ensembles, first-order Markov linear copolymer ensembles, nonlinear star polymer ensembles with varying arm-length, topology and composition, and polymer ensembles represented by molecular mass distributions. These examples demonstrate that the earth mover's distance captures differences neglected by the average method and offers greater resolutions of chemical distinctions between polymer ensembles. With no supervision, the use of earth mover's distance metric gives a quantitative and reliable numeric calculation of pairwise similarity between two polymer ensembles. [Shi et al. ACS Polymers Au 2024, 4, 1, 66–76]

Our similarity methods can handle sequential, compositional, molecular mass, and topological differences, which are typically either entirely or partially ignored by traditional methods, enabling distinctions between different ensembles that would otherwise be erroneously determined to be identical. In conclusion, our similarity methodology represents a critical advancement in the quantitative calculation of polymer similarity and accelerates the progress of cheminformatics for polymers including for applications such as property prediction and classification.

169cb Development of Chimes Machine Learned Interatomic Potentials for All Silica MFI Zeolites

Vallabh Vasudevan, Sayed Ahmad Almohri, Anakin Bahm, Rebecca Lindsey

The Haber-Bosch (HB) process has revolutionized agricultural output to match the ever-growing global food supply demand by enabling mass production of ammonia, a key precursor in production of synthetic fertilizers. The HB process works by combining nitrogen and hydrogen at high temperatures and pressures in the presence of catalysts, causing the entire process to be highly energy intensive. The process used to separate the ammonia from the gaseous reactive mixture is particularly energy intensive, requiring the gas to be cooled to 240 K to liquefy the ammonia. The current strategy for reducing energy requirements is two pronged: developing better catalysts to drive the reaction equilibrium and materials for enhanced separation efficiency.

Zeolites membranes are naturally occurring nanoporous structures of aluminosilicate compounds which show great promise as separation medium, exhibiting a near continuum of pore sizes that allow for separation of target gas molecules. Zeolite membranes have high thermal and chemical stabilities and show good selectivity and permeability for the separation of various mixtures. [1–4] Key factors controlling performance of the zeolite membrane are spatial variation in adsorption, diffusion pathways and pore channel length, all of which are affected by the atomic arrangement of the zeolite framework. [5] Experimental techniques have been used to probe these features with atomic level resolution to image zeolite atomic arrangements. [6,7] The understanding of the atomic arrangement can provide some key insights into the possible channel and diffusion dynamics. An extensive understanding of the molecular mechanisms affecting adsorption, diffusion and selectivity and dynamic behavior of the framework are hard to approach using current experimental techniques. [5–7] Here, we present a new machine learned interatomic model for bulk MFI zeolite using the ChIMES physics-informed machine learned interatomic model and development framework. [8–14] We will show that bulk structural properties predicted with our model such as radial distribution functions and vibrational power spectra compare well with similar properties derived from DFT simulations. Our newly developed ChIMES machine learned interatomic potential for all-silica MFI zeolite can be used to provide useful insight into the role of zeolite as a separation media for Haber-Bosch process. Our model can be used to simulate the zeolite pore channels and compare the predicted structural configurations with observed experimental structures derived from imaging techniques such as ptychography to better inform and predict atomic level framework dynamics and performance characteristics. [5] We will discuss a possible strategies to extend our MFI zeolite model to other all-silica zeolite structures. Ultimately, our newly developed ChIMES machine learned IAP for all-silica MFI zeolite can be used to provide useful insight into the role of zeolite as a separation media for Haber-Bosch process and to study the underlying interactions in zeolite-mediated catalytic processes.

References

[1] Hisao Inami et al. Membranes 11.5 (2021)

[2] Miguel Palomino et al. Langmuir 26.3 (2010)

[3] Jin Shang et al. Chem. Commun. 50 (35 2014)

[4] Jin Shang et al. J. Am. Chem. Soc. 134.46 (2012)

[5] Hui Zhang et al. Science 380.6645 (2023), pp. 633–638.

[6] Zhuoya Dong et al. J. Am. Chem. Soc. 145.12 (2023).

[7] Haozhi Sha et al. Science Advances 9.11 (2023)

[8] Rebecca K. Lindsey et al. J. Chem. Theory Comput. 13.12 (Dec. 2017).

[9] Rebecca K. Lindsey et al. The Journal of Chemical Physics 153.13 (2020)

[10] Rebecca K. Lindsey et al. J. Chem. Theory Comput. 15.1 (2019)

[12] ChIMES Calculator. URL: https://github.com/rk-lindsey/chimes_calculator.

[13] ChIMES Generator. URL: https://github.com/rk-lindsey/chimes_lsq.

[14] ChIMES Active Learning Driver. URL: https://github.com/rk-lindsey/al_driver.

169cc Development of Experimental Guidelines for Organic Field-Effect Transistors (OFETs) Using Machine Learning Based on Ofets Database

Myeongyeon Lee, Rahul Venkatesh, Jessica A. Bonsu, Carson Meredith, Martha Grover, Elsa Reichmanis

Conjugated polymers are attractive materials for flexible and stretchable electronic device applications with cost-effective processes. Although their performance is not expected to reach that of silicon-based devices, they can be applied to large area and curved surfaces. However, the device fabrication steps include many process variables that consume abundant resources and time to optimize the process with trial-and-error experimentation. In the case of organic field-effect transistors (OFETs), trial-and-error experimentation has been employed to explore optimization in huge process space. This bottleneck can be overcome with data-driven approaches. In this study, we propose a database structure to store the process conditions and guide future experimental directions More than 600 data points poly(3-hexylthiophene) (P3HT), poly[2,5-(2-octyldodecyl)-3,6-diketopyrrolopyrrole-alt-5,5-(2,5-di(thien-2-yl)thieno [3,2-b]thiophene)] (DPP-DTT), and poly{[N,N'-bis(2-octyldodecyl)naphthalene-1,4,5,8-bis(dicarboximide)-2,6-diyl]-alt-5,5'-(2,2'-bithiophene)} (N2200) are extracted and machine learning is used to understand the relationship between each process variable and device performance. We identified which variable had a significant impact on the device performance. The study shows that experimental guidelines can be provided by machine learning to reduce resources and time for process optimization.

169cf Molecular Dynamics Simulation Study of the Huntingtin Fibril's Morphology, Stability, Kinetics, and Role of Water

Avijeet Kulshrestha, Azamat Rizuan, Jeetain Mittal

Protein aggregation during aging is implicated in numerous incurable neurodegenerative diseases, wherein functional proteins transform into dysfunctional biological condensates or, in some cases, create pathological amyloid fibrils through a liquid-to-solid transition (LST). The repetition of glutamine residues has a tendency to form pathological amyloid fibrils and has been linked to Polyglutamine (PolyQ) diseases, such as spinocerebellar ataxia, dentatorubral-pallidoluysian atrophy, Huntington's disease, and spinal-bulbar muscular atrophy. Despite ongoing research efforts, there has been only indirect experimental measurement of the structure of PolyQ fibrils, and a comprehensive understanding of the early and later stages of amyloid formation remains elusive. In our study, we employ a multiscale simulation framework to test two available fibril structures of PolyQ, i.e., -arc and -turn. Although the -arc model shows stronger fibril contacts, the -turn model exhibits a higher stability. This attribute is associated with the observed higher interaction of water molecules within the -turn fibril core, producing a hydrogen bonding network. We further investigate the effects of temperature and concentration on the mechanisms of fibrillation and fibril morphology. Our findings uncover a high degree of heterogeneity in fibril morphology, with distinct morphologies associated with varying levels of toxicity, and reveal that the antiparallel -turn model of the PolyQ fibril exhibits faster kinetics.

169ch The Hidden Chokepoints: Exploring Gas Diffusion in the Codh/ACS Enzyme Complex Using Molecular Simulations.

Suman Samantray, Bojana Ginovska-Pangovska, Simone Raugei

The carbon monoxide dehydrogenase (CODH) and acetyl-CoA synthase (ACS) enzymatic complex catalyzes the (reversible) electrochemical conversion of CO_2 to CO and the subsequent condensation of CO with a methyl intermediate and coenzyme A to form acetyl coenzyme A. CO_2 reduction happens at the CODH C-cluster, and the CO product is transported and consumed at the ACS A-cluster. The bifunctionality of the complex is facilitated by a 140 Å-long gas channel for the transport of CO between the C- and A-clusters.

CODH and the CODH/ACS are central for atmospheric carbon monoxide oxidation and supporting microbial life. These enzymes are also the focus of recent biotechnological applications because of their potential replacement of two energy-intensive industrial processes: 1) the water–gas shift reaction, a widely used industrial process for H_2 or CO production, and 2) the Monsanto process used to produce acetic acid from methanol and CO. However, the effective and widespread use of CODH and CODH/ACS for industrial fields is limited by their poor activity under aerobic conditions. One strategy to design O_2-tolerant complexes is to prevent O_2 from entering the enzyme matrix.

In this work, we use atomistic free energy simulations to explore the gas diffusion through the CODH/ACS complex from M. thermoacetica and other design variants that favor uptake of CO_2 and O_2. We also explore how the transport mechanism is affected by conformational changes in the subunits of the enzyme caused due to transformation of cavities and gating mechanism of residues.

Research Interests:

"Peptoids" as a viable biomaterial in food packaging industry.

Antifreeze proteins (AFPs) stand as a paradigm-shifting innovation, that prevent freezing in organisms living in sub-zero environments by inhibiting ice crystallization. AFPs, such as Type I AFP from winter flounder and Type III AFP from ocean pout, have been instrumental in preventing ice crystallization in biological and industrial settings. This unique function has profound implications in cryopreservation, organ transplant technology, and notably in the food industry, where AFPs play a pivotal role in enhancing product longevity and quality.

Peptoids are synthetically engineered polymers, that could mimic natural AFPs with enhanced flexibility. The fundamental distinction between peptoids and peptides, while structurally similar, differ in the placement of their side chains. In peptides, side chains are attached to the alpha carbon atoms, whereas in peptoids, they are bonded to the nitrogen atoms of peptide backbone. Peptoids could be customized with targeted functional properties, to increase stability against enzymatic degradation, higher resilience under various physical conditions, and a reduced risk of immune responses, making them more suitable for industrial applications.

The research will address critical scientific questions: How do structural variations in peptoids influence their binding efficiency and antifreeze activity compared to natural AFPs? What thermodynamic and kinetic parameters are crucial in determining the antifreeze efficiency of peptoids? How can molecular dynamics simulations contribute to the design of optimized peptoid AFPs for industrial applications, particularly in food packaging?

Circular economy: Conversion of biomass-plastic waste to biochar-based adsorbents.

The crisis of plastic waste, particularly the accumulation of microplastics in the environment, presents a critical challenge for sustainable development. Addressing this issue requires innovative solutions that not only decompose plastic waste but also repurpose it into beneficial materials. Artificial metalloenzymes (ArMs) offer a targeted approach to catalyze the degradation of resistant plastic polymers, transforming them into value-added products such as biochar. Biochar, derived from the carbonization of organic matter, is recognized for its adsorptive properties, and is proposed mechanism to sequester pollutants.

Natural metalloenzymes like laccase and manganese peroxidase catalyze essential reactions in biological systems, including the degradation of complex molecules. Mimicking natural enzymes, artificial metalloenzymes can be engineered to target the sturdy chemical bonds in plastics such as polyethylene terephthalate (PET), high-density polyethylene (HDPE), and polystyrene (PS). Advanced molecular simulation techniques coupled with rational design and directed evolution can finely tune the active sites of metalloenzymes for specific substrate interaction, enhance catalytic rates, and provide stability against harsh industrial conditions.

The research aims to: Design artificial metalloenzymes with enhanced specificity for breaking down multiple types of plastics, focusing on the creation of intermediates destined for biochar conversion. Explore metabolic engineering of microorganisms to host these ArMs, evaluate the life-cycle of the enzymatic byproducts, and upscale the entire process for industrial use. Successful implementation would represent a dual achievement: mitigating plastic pollution and harnessing the resulting biochar for environmental adsorption applications, thereby fostering a circular economy in waste management.

PFAS-mediated dysregulation of biomolecular condensates.

The pervasive presence of per- and polyfluoroalkyl substances (PFAS) in the environment, coupled with their long biological half-life, has emerged as a concern for public health. PFAS compounds like PFOA and PFOS, notorious for their toxicity, exposure to it is linked to a multitude of health risks, and there is a critical need to understand their interaction with cellular components at a molecular level. This research focuses on the impact of PFAS on biomolecular condensates, which are crucial for various cellular processes, including RNA metabolism, signal transduction, and the stress response.

Biomolecular condensates, formed via liquid-liquid phase separation, are highly dynamic and sensitive to changes in their microenvironment. The physicochemical properties of PFAS, such as their hydrophobicity and stability, make them potential disruptors of the condensates' structural integrity. PFAS affect biomolecular condensates at the molecular level, this research aims to utilize molecular simulations to bridge the gap between environmental exposure and cellular dysfunction, offering a foundational perspective on the health implications of PFAS. This work has the potential to inform risk assessments, shape regulatory decisions, and inspire the development of therapeutic interventions to counteract the detrimental effects of PFAS on human health.

Teaching Interests:

The beauty of academic life is that it gives an opportunity to learn, teach, and work with young, bright, and dynamic students, who are the future of our society. Through the influences of the great teachers, I am at this stage to pursue academic life as a part of devoted teaching and research to serve our society. In the following, I describe in detail experiences, teaching philosophy, and plan for teaching.

Teaching experience: I have had a privilege of being TA for the computational biotechnology practical courses of MSc students at University of Galway, Ireland and RWTH Aachen University, Germany. In addition, I co-mentored theses of BSc, MSc students, and students.

Teaching Philosophy: In my opinion, the most important aspect of teaching is that we learn, teach, work, and develop together for the welfare and service of our society. I believe that we can learn best by 'doing'. The fundamental aspect of teaching is igniting the mind of students with the curiosity and skills necessary to develop questions as well as guide them one-step further to motivate, promote and develop skills and scientific processes necessary to investigate the answers to those questions. To facilitate this, I will keep my classes interactive and open for discussion. Besides, I will take the opportunity to discuss teaching, learning, training, and research with students, friends, colleagues, senior professors, and experts for the betterment of my service to the institute and society.

Teaching Plan: I am interested in teaching the core and professional courses of Biochemical Engineering. In the core courses, I am in the teaching of Protein thermodynamics, Statistical mechanics, Chemical kinetics, and Chemical reaction Engineering. In the professional courses, I would like to teach Bioinformatics, Structural Biology, and Biophysics. Besides, I have the interest to design courses, e.g., Bioeconomy, Protein Engineering.

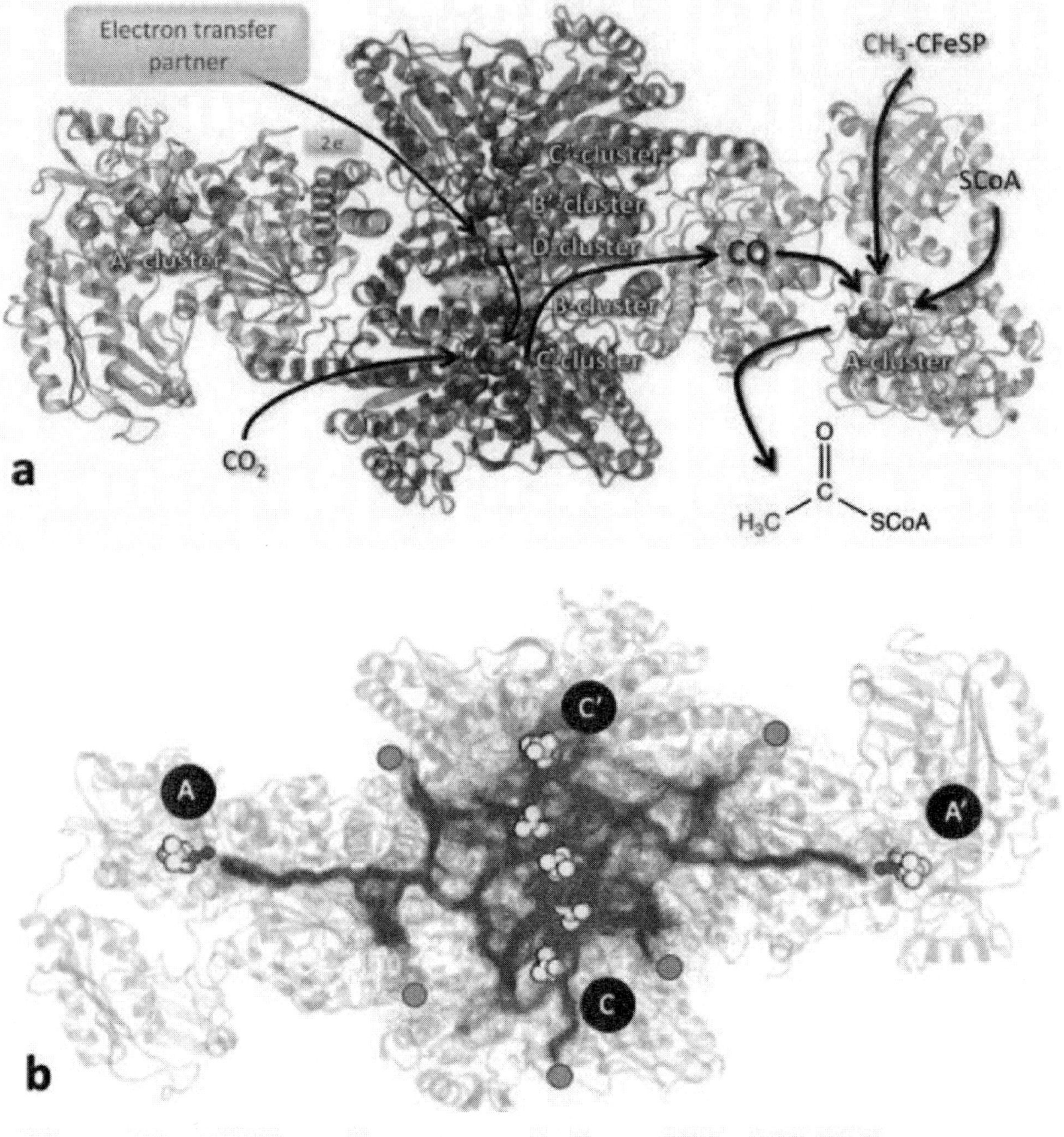

Figure 1. a) Schematic representation of the CO_2/CO processes catalyzed by the CODH/ACS complex from *M. thermoacetica*, with the active metalloclusters A- and C- labelled in cyan. b) A composite view of gas channels (dark gray) in CODH/ACS discovered by long-timescale molecular dynamics simulation.

169cj Molecular Design of High Performance Electrolytes with Generative Algorithm

Vignesh Sathyaseelan, Brett Savoie

The quest for advanced battery technologies continues to drive innovation in the field of energy storage. Fluoride-ion batteries (FIBs) are a novel battery chemistry that has the potential to provide significant improvements over existing lithium-ion batteries due to its high charge capacity and low volumetric changes during intercalation; as well as the abundance and low cost of raw material required to manufacture them. However, the successful implementation of fluoride-ion batteries hinges on overcoming challenges associated with the design of stable solvents for FIB electrolytes that facilitate efficient ion transport while maintaining chemical stability over numerous charge-discharge cycles at room temperature operations. In this endeavor, the application of genetic algorithms presents a cutting-edge approach, offering a powerful means to explore the vast chemical space and design novel electrolytes for fluoride-ion batteries.

In recent years, the field of computational chemistry has witnessed a surge in novel approaches aimed at designing novel materials for energy storage applications. One such technique involves harnessing the power of Genetic Algorithms (GA) to optimize molecular structure for target properties by evolving and selecting candidate molecules. The goal of this study is to design solvents for FIB electrolytes that concurrently exhibit robust chemical stability against fluoride-ion attacks and facilitate efficient ion transport, making them promising candidates for use in fluoride ion batteries using a machine learning augmented genetic algorithm.

The molecular optimization process is guided by carefully crafted fitness functions that encapsulate the key criteria for stability against fluoride ion attacks as well as the ability of the solvent to solvate fluoride ions. This loss function incorporates two critical components: a stability score and a solvation score. The stability score (optimized using GA to be higher) is derived from the application of the Bell–Evans–Polanyi (BEP) principle, which allows for the evaluation of activation barrier trends by computing the heat of a potential reaction using the GFN2-xTB method. Enumeration of all conceivable products arising from the reaction of a molecule with fluoride ion is executed using its chemical graph, from which the lowest activation barrier is determined based on this stability score. This principle serves as a robust metric in predicting the likelihood of a molecular structure remaining stable under the influence of fluoride ions.

Concurrently, the solvation score plays a crucial role in the molecular optimization process. Departing from conventional methodologies that rely on computationally expensive simulations, the solvation score (optimized using GA to be lower) in this approach leverages machine learning techniques to predict solvation-free energy efficiently. We train a random forest (RF) model to predict the solvation-free energy of a designed solvent molecule for fluoride ions. Inexpensive molecular features such as Morgan Fingerprint and Solvent Accessible Surface Area (SASA) were calculated which serves as input featurization to this RF model. RF models were trained and tested on an 80:20 split of a dataset consisting of 643 data points in the model. This machine learning model streamlines the computational process but also enables the rapid assessment of the solvation effects of the designed molecules.

Utilizing this framework, potential molecular candidates were generated and subjected to rigorous higher-level evaluation to study their viability as solvents for electrolytes in FIBs. This evaluation entailed characterization of the reaction network with fluoride ions, alongside the assessment of their efficacy in solvating fluoride ions, employing higher-level Density Functional Theory (DFT) calculations.
In summary, the integration of genetic algorithms and theoretical principles for multi-objective optimization in the chemical space showcases a promising avenue for designing stable molecules tailored for fluoride ion batteries.

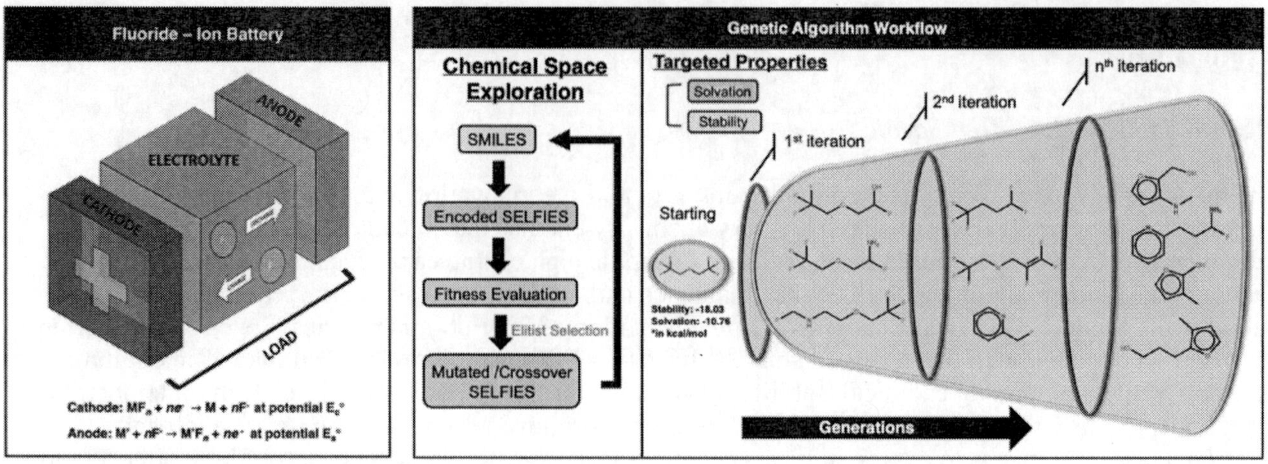
Fluoride – Ion Battery
ANODE
ELECTROLYTE
CATHODE
LOAD
Cathode: MFn + ne⁻ → M + nF⁻ at potential Ec°
Anode: M' + nF⁻ → M'Fn + ne⁻ at potential Ea°
Genetic Algorithm Workflow
Chemical Space Exploration
SMILES
Encoded SELFIES
Fitness Evaluation
Elitist Selection
Mutated /Crossover SELFIES
Targeted Properties
Solvation
Stability
1st iteration
2nd iteration
nth iteration
Starting
Stability: -18.03
Solvation: -10.76
*in kcal/mol
Generations

169cl Generative Artificial Intelligence for Property-Guided Design of Co-Polymers

Gabriel Vogel, Jana Weber

Synthetic polymers are an important class of materials that one finds in many different consumer products, ranging from energy materials, over plastics, to medical applications. There is a constant demand for novel polymer materials with improved highly specialised functionalities. Traditional efforts in discovery and development of novel materials are often time-consuming, require much expert knowledge, and do not consider the entire possible design space. One promising step towards more efficient, while at the same time broader, exploration of the chemical polymer design space is the development of generative molecular machine learning models.

Generative artificial intelligence (AI) for the design of synthetic polymers still needs to overcome domain-specific challenges. One challenge is that unlike for small molecules, synthetic polymers are governed by multiple structural levels of information that go beyond the atomic structure of monomers. Polymers are governed by monomer stoichiometries, by chain architectures, linking structures, chain length and many more. This raises the question on how to best represent polymers for machine learning algorithms. Secondly, controlled design of novel materials necessitates much (property-) labelled data, which in the field of synthetic polymers is not yet easily accessible.

We present our currently developed approaches on molecular machine learning for co-polymer design. We build upon the representation of polymers as molecular graph ensembles [1] and work on the two challenges outlined above: learning with limited labelled data and learning beyond the atomic representation of monomer units. To this end, we extend our previous Graph-to-string variation autoencoder [2] to take partly labelled-data into account and to organise its latent space based on target property information. We additionally perform Bayesian optimisation in the models latent space to identify most relevant property regions.

We illustrate our approach on a case study for polymer photocatalyst design for the production of green hydrogen. Our results show that we can sample 100% valid co-polymers with high novelty (> 90%) and diversity (> 80%) from desired property regions in the latent space [3]. Notably our model extends previous generative AI works on polymers by designing for the first time molecular ensembles; therewith including stoichiometries and chain architectures. This presents an important step in the direction of generative AI for structurally diverse material classes.

References

[1]: Aldeghi, M., & Coley, C. W. (2022). A graph representation of molecular ensembles for polymer property prediction. Chemical Science, 13(35), 10486-10498.

[2]: Vogel, G., Sortino, P., & Weber, J. M. (2023). Graph-to-String Variational Autoencoder for Synthetic Polymer Design. In *AI for Accelerated Materials Design-NeurIPS 2023 Workshop.*

[3]: Vogel, G., Weber, J.M. (*In preparation*). Inverse design of copolymers with optimized properties.

169cm Matlab Data Processing and AI for Molecular Chemistry

Hossein Jooya, Ramnarayan Krishnamurthy, Kathi Kugler, Aycan Hacioglu

We will show how various types of chemical data are imported, processed, and analyzed in MATLAB. This includes NMR spectra, Cryo-EM images, and molecular structures. We will cover how the integration with RDKit allows us to prepare large chemical data for different machine learning and deep learning tasks. This will be demonstrated as a workflow that integrates deep learning QSAR predictions into PKPD models in MATLAB.

169cn Accelerating Drug Discovery through the Automatic Population of a Pharmaceutical Ontology Using Knowledge Graphs

Naz Pinar Taskiran, Arijit Chakraborty, Vipul Mann, Venkat Venkatasubramanian

In the process of drug discovery and manufacturing, experts need to scan through hundreds to thousands of documents to obtain relevant information. One way of overcoming the challenges posed by this process is the use of ontologies, which model domain knowledge hierarchically. Currently, the field lacks domain-specific, automatically populated ontologies that store large amounts of data. To address this, we developed SUSIE, an ontology-based pharmaceutical information extraction tool built to extract semantic triples[1]. These are presented to the user in the form of knowledge graphs (KGs). The Columbia Ontology of Pharmaceutical Engineering (COPE) is used to extract the semantic triples, but the KGs do not match its hierarchical structure[2]. We present a novel framework for automatically populating the sub-classes of COPE. By its hierarchical nature, this also allows us to infer the other classes that contain the extracted terms. In addition to expanding SUSIE's training base, fine-tuning the entity recognition process, and expanding the ontology classes, automatic mapping of KG nodes and edges to COPE allowed us to create a database of logically organized pharmaceutical terms, which aims to streamline the querying process and accelerate the drug discovery and manufacturing pipeline.

Bibliography:

(1) Mann V., Viswanath S., Vaidyaraman S., Balakrishnan J., Venkatasubramanian V., (2023).

SUSIE: Pharmaceutical CMC ontology-based information extraction for drug development using machine learning,

Computers & Chemical Engineering, Volume 179, 108446.

https://doi.org/10.1016/j.compchemeng.2023.108446.

(2) Remolona M. F. M., Conway M. F. , Balasubramanian S., Fan L., Feng Z., Gu T., Kim H., Nirantar P. M. , Panda S., Ranabothu N. R. , Rastogi N., Venkatasubramanian V., (2017).

Hybrid ontology-learning materials engineering system for pharmaceutical products: Multi-label entity recognition and concept detection.

Computers & Chemical Engineering, Volume 107, Pages 49-60.

https://doi.org/10.1016/j.compchemeng.2017.03.012.

169co A Based Take on Bayesian Optimization: Tuning Kernelized Bandits for Expensive Experiments with Mixed, Discrete Inputs

Christoph Griehl, Kai Sundmacher

In recent years, Bayesian Optimization (BO), also recognized as kernelized bandits, has emerged as a powerful tool, transcending the boundaries of traditional informatics to find successful applications in diverse fields such as drug discovery, material design, and chemical synthesis [1], [2], [3], [4]. This widespread adoption is largely due to the method's non-parametric nature and its efficiency regarding the number of evaluations.

The development of numerous extensions has facilitated the application of BO across complex, high-dimensional, and mixed search spaces [5], [6], [7], [8]. Moreover, there has been a concerted effort towards the effective implementation of BO, particularly for automating closed-loop High Throughput Experimentation processes [9], [10], [11], [12]. However, the adaptation of kernelized bandits to specifically guide laboratory experimentation and manage resource-intensive simulations has received relatively scant attention. Traditional BO implementations often seek to balance the number of iterations against computational time. Yet, in the realm of laboratory experiments and expensive simulations, prioritizing a reduction in iteration count—even at the cost of increased computational time—can offer substantial benefits.

Additionally, while there has been progress in optimizing mixed search spaces that incorporate categorical, ordinal, or binary inputs [5], [6], [7], [8], the treatment of discrete numerical inputs has received comparatively less attention. The growing accessibility and utilization of chemical descriptors, encodings, or embeddings [13], [14], coupled with the constraints imposed by some laboratory setups on searching over a fully continuous parameter space, necessitate a refined approach for this category of inputs.

This congress contribution explores various methodologies, namely probabilistic reparameterization, interleaved optimization and continuous relaxation, for optimizing the acquisition function in search spaces that include discrete numerical inputs. This step is considered crucial for adapting traditional BO approaches to non-continuous inputs [15], [16]. Additionally, we are testing a novel type-dependent embedding strategy that combines random matrix embeddings for continuous inputs [6], [17], [18] with traditional, bi-directional methods for complexity reduction, specifically, Principal Component Analysis, in discrete properties. This approach facilitates application in high-dimensional search spaces, which is particularly pertinent when selecting a large array of descriptors to represent each class of reaction agents.

Despite BO's non-parametric status, the workflow of chemical engineers and chemists encompasses numerous considerations, such as the original encoding or embedding of chemical species and the initialization strategy, which significantly impact the optimization process's efficacy [3]. Our research also delves into the influence of these factors.

An algorithm, informed by our findings and benchmarked against prevailing BO strategies in a series of experiments within chemical engineering and synthesis, has been developed. The algorithm's open-source codebase is constructed using the PyTorch [19] family, enabling seamless integration with established frameworks such as BoTorch [20] and GPyTorch [21].

This contribution aims not only to advance the theoretical understanding and practical application of BO in guiding laboratory experimentation and simulations but also to foster the development of more efficient and effective optimization strategies in the chemical sciences.

References

[1] J. A. G. Torres *et al.*, 'A Multi-Objective Active Learning Platform and Web App for Reaction Optimization', *J. Am. Chem. Soc.*, vol. 144, no. 43, pp. 19999–20007, Nov. 2022, doi:

10.1021/jacs.2c08592.

[2] K. Wang and A. W. Dowling, 'Bayesian optimization for chemical products and functional materials', *Current Opinion in Chemical Engineering*, vol. 36, p. 100728, Jun. 2022, doi: 10.1016/j.coche.2021.100728.

[3] B. J. Shields *et al.*, 'Bayesian reaction optimization as a tool for chemical synthesis', *Nature*, vol. 590, no. 7844, pp. 89–96, Feb. 2021, doi: 10.1038/s41586-021-03213-y.

[4] F. Häse, L. M. Roch, and A. Aspuru-Guzik, 'Chimera: enabling hierarchy based multi-objective optimization for self-driving laboratories', *Chem. Sci.*, vol. 9, no. 39, pp. 7642–7655, 2018, doi: 10.1039/C8SC02239A.

[5] X. Wan, V. Nguyen, H. Ha, B. Ru, C. Lu, and M. A. Osborne, 'Think Global and Act Local: Bayesian Optimisation over High-Dimensional Categorical and Mixed Search Spaces'. arXiv, Jun. 10, 2021. Accessed: Nov. 16, 2023. [Online]. Available: http://arxiv.org/abs/2102.07188

[6] L. Papenmeier, L. Nardi, and M. Poloczek, 'Bounce: a Reliable Bayesian Optimization Algorithm for Combinatorial and Mixed Spaces'. arXiv, Jul. 02, 2023. Accessed: Aug. 25, 2023. [Online]. Available: http://arxiv.org/abs/2307.00618

[7] C. Oh, J. M. Tomczak, E. Gavves, and M. Welling, 'Combinatorial Bayesian Optimization using the Graph Cartesian Product'. arXiv, Oct. 28, 2019. Accessed: Apr. 07, 2024. [Online]. Available: http://arxiv.org/abs/1902.00448

[8] A. Deshwal, S. Ament, M. Balandat, E. Bakshy, J. R. Doppa, and D. Eriksson, 'Bayesian Optimization over High-Dimensional Combinatorial Spaces via Dictionary-based Embeddings', Mar. 2023.

[9] O. J. Kershaw *et al.*, 'Machine learning directed multi-objective optimization of mixed variable chemical systems', *Chemical Engineering Journal*, vol. 451, p. 138443, Jan. 2023, doi: 10.1016/j.cej.2022.138443.

[10] A. M. K. Nambiar, C. P. Breen, T. Hart, T. Kulesza, T. F. Jamison, and K. F. Jensen, 'Bayesian Optimization of Computer-Proposed Multistep Synthetic Routes on an Automated Robotic Flow Platform', *ACS Cent. Sci.*, vol. 8, no. 6, pp. 825–836, Jun. 2022, doi: 10.1021/acscentsci.2c00207.

[11] M. Christensen *et al.*, 'Data-science driven autonomous process optimization', *Commun Chem*, vol. 4, no. 1, p. 112, Dec. 2021, doi: 10.1038/s42004-021-00550-x.

[12] A. E. Gongora *et al.*, 'A Bayesian experimental autonomous researcher for mechanical design', *Sci. Adv.*, vol. 6, no. 15, p. eaaz1708, Apr. 2020, doi: 10.1126/sciadv.aaz1708.

[13] T. Gensch *et al.*, 'A Comprehensive Discovery Platform for Organophosphorus Ligands for Catalysis', *J. Am. Chem. Soc.*, vol. 144, no. 3, pp. 1205–1217, Jan. 2022, doi: 10.1021/jacs.1c09718.

[14] J. Ross, B. Belgodere, V. Chenthamarakshan, I. Padhi, Y. Mroueh, and P. Das, 'Large-Scale Chemical Language Representations Capture Molecular Structure and Properties'. arXiv, Dec. 14, 2022. Accessed: Apr. 07, 2024. [Online]. Available: http://arxiv.org/abs/2106.09553

[15] E. C. Garrido-Merchán and D. Hernández-Lobato, 'Dealing with Categorical and Integer-valued Variables in Bayesian Optimization with Gaussian Processes', *Neurocomputing*, vol. 380, pp. 20–35, Mar. 2020, doi: 10.1016/j.neucom.2019.11.004.

[16] S. Daulton, X. Wan, D. Eriksson, M. Balandat, M. A. Osborne, and E. Bakshy, 'Bayesian Optimization over Discrete and Mixed Spaces via Probabilistic Reparameterization'. arXiv, Oct. 18, 2022. Accessed: Oct. 18, 2023. [Online]. Available: http://arxiv.org/abs/2210.10199

[17] J. Kirschner, M. Mutný, N. Hiller, R. Ischebeck, and A. Krause, 'Adaptive and Safe Bayesian Optimization in High Dimensions via One-Dimensional Subspaces'. arXiv, May 28, 2019. Accessed: Apr. 08, 2023. [Online]. Available: http://arxiv.org/abs/1902.03229

[18] A. Munteanu, A. Nayebi, and M. Poloczek, 'A Framework for Bayesian Optimization in Embedded Subspaces', 2019.

[19] P. Wu, 'PyTorch 2.0: The Journey to Bringing Compiler Technologies to the Core of PyTorch (Keynote)', in *Proceedings of the 21st ACM/IEEE International Symposium on Code Generation and Optimization*, Montréal QC Canada: ACM, Feb. 2023, pp. 1–1. doi: 10.1145/3579990.3583093.

[20] M. Balandat *et al.*, 'BOTORCH: A Framework for Efficient Monte-Carlo Bayesian Optimization', *Advances in Neural Information Processing Systems*, vol. 33, Dec. 2020.

[21] J. Gardner, G. Pleiss, K. Q. Weinberger, D. Bindel, and A. G. Wilson, 'GPyTorch: Blackbox Matrix-Matrix Gaussian Process Inference with GPU Acceleration'.

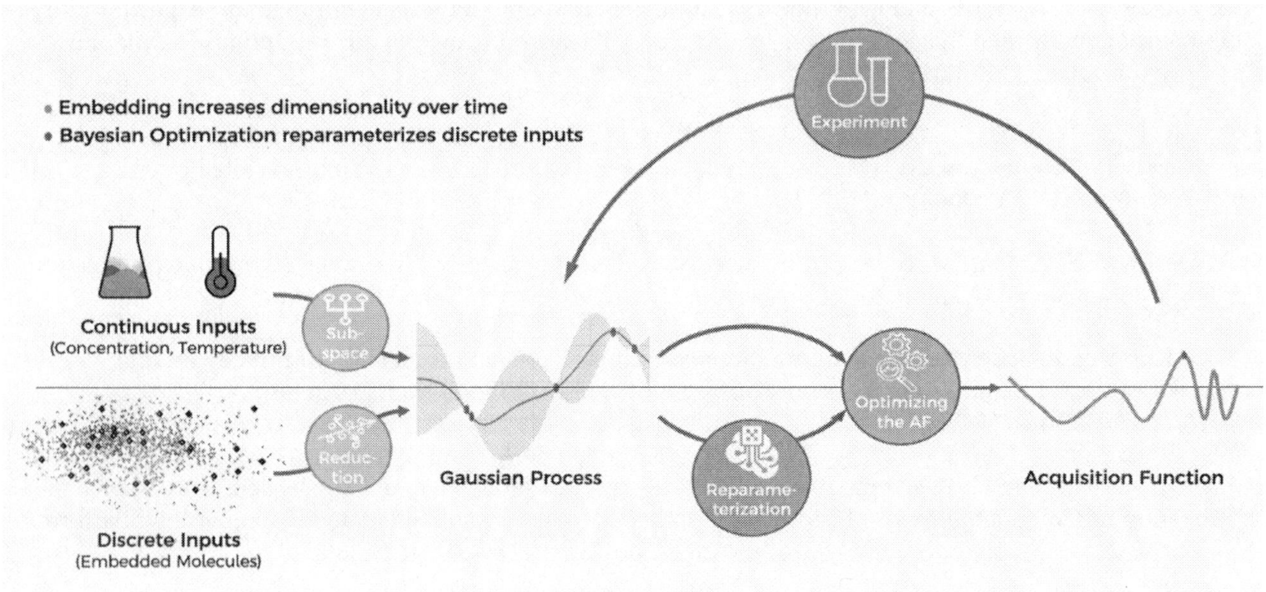

169cq Advancing Molecular Property Prediction and Chemical Reactivity Understanding through High-Throughput Quantum Chemistry and Graph Neural Networks

Haoyang Wu, Hao-Wei Pang, Jonathan Zheng, Jackson W. Burns, Xiaorui Dong, Shih-Cheng Li, Sayandeep Biswas, Kevin Spiekermann, Angiras Menon, William Green

In this talk, we present our work in generating a one-of-a-kind foundational quantum mechanical (QM) dataset for thermochemistry, reaction kinetics, solvation free energies, and QM descriptors, generated using our open-source high-throughput computational chemistry workflow. This comprehensive dataset comprises over 345,000 open- and closed-shell reacting species, 200,000 transition states optimized at both the GFN2-XTB semi-empirical and ωB97X-D/def2-SVP density functional theory (DFT) levels of theory, DLPNO-CCSD(T)-F12d/def2-TZVP reaction barrier heights, and more than 100 million COSMO-RS solvation free energies in over 300 common solvents. The unparalleled scale and depth of this dataset can potentially lead to unprecedented insights into molecular properties and chemical reactivity.

By leveraging this extensive dataset and state-of-the-art Chemprop Directed Message Passing Neural Network (D-MPNN) models, we present promising preliminary results in accurately predicting molecular properties and chemical reactivity, such as the oxidative stability of diverse materials and solvent effects on reactivity. The generated dataset also paves the way for training foundational models, exploring active learning methods, and developing reactive neural network force fields. These advancements have far-reaching implications for further advancing autonomous chemical and material design, accelerating the discovery of novel materials with desired functional properties.

The combination of high-throughput quantum chemistry and advanced graph neural networks demonstrates the immense potential for accelerating materials discovery and optimization. By harnessing the power of this unique dataset and cutting-edge machine learning techniques, we pave the way for a deeper understanding of molecular properties and chemical reactivity, ultimately leading to more efficient and targeted design of functional materials for a wide range of applications.

169cs Advancing Molecular Screening: Integrating Operating Conditions into Transformer Models for Chemical Engineering

Aahil Khambhawala, Silabrata Pahari, Chi-Ho Lee, Joseph Kwon

The chemical industry has undergone considerable advancements, yet a pressing requirement for further development of processes to reduce energy consumption, increase yield, and meet production demands persists. Engineers have multiple strategies at their disposal to boost the efficiency of processes. Among these, process optimization through intensification[1], heat integration[2], and the advancement of chemicals used as solvents, catalysts, and promoters are particularly significant. The discovery of new chemicals is the most direct route to substantial enhancements within these approaches. Nonetheless, the experimental discovery of new solvents proves to be an exceedingly demanding task.

An alternative to experimental approaches is utilizing computational methods such as density functional theory and molecular dynamics. These techniques considerably alleviate the arduous process of conducting experiments. However, they are often associated with high computational costs and lengthy time requirements. Various machine learning-based methodologies have been adopted to surmount this challenge, owing to their reduced computational demands and rapid prediction capabilities[3][4]. These methods excel in making system-specific predictions and have advanced the development of rapid quantitative structure-property relationships. Nonetheless, traditional machine-learning approaches are limited in generalizing across different chemical systems. They frequently necessitate extensive feature engineering and are typically tailored for a limited range of compounds, restricting their applicability to other molecules. In contrast, recent progress has been made with transformer-based machine learning models. These models have demonstrated potential in cheminformatics, providing accurate predictions of system-specific properties using the SMILES notation of compounds. This represents a notable enhancement over previous models, primarily due to the self-attention mechanism, which facilitates a more profound understanding of intermolecular interactions. However, despite the significant advances in employing transformers to predict properties under standard conditions effectively, their application in forecasting properties under diverse conditions, essential for the design and optimization of industrial processes, remains an area in need of further exploration and development.

In our research, we have incorporated various operating conditions characteristic of chemical processes to enhance the molecular screening process, thereby aiding engineers in developing optimized processes. We employed a transformer model pre-trained on an extensive dataset comprising 1.6 billion molecules from the PubChem database. This pre-training enables the model to capture the complex relationships between chemical structures and their properties. The encoder within the transformer generates a numeric representation of molecules, which can be regarded as a 'universal chemical language' for property prediction. During the fine-tuning stage, the molecules are processed through the encoder to obtain molecular embeddings. These embeddings are subsequently combined with the up-projected operating conditions and input into a neural network with residual connections to predict the target property. Incorporating residual connections allows the neural network to retain the molecular correlations obtained from the encoder while giving equal consideration to the operating conditions introduced later. This approach addresses two significant challenges in utilizing transformers for property predictions in chemical engineering. Firstly, we obviate the need for a large labeled dataset to train the encoder by excluding operating conditions from the pre-training phase. This is particularly beneficial given the scarcity of high-quality data in many chemical engineering domains, which has previously impeded the application of transformers in these areas. Secondly, we augment the model's relevance for process design by incorporating operating condition augmentations. Given the dynamic nature of chemical processes, the ability to predict properties under varying conditions is essential for designing more efficient processes. This methodology enables its application in a broad spectrum of chemical systems for purposes including, but not limited to, catalyst discovery, drug development, and solvent exploration from a process design perspective. Our architecture minimizes the computational burden of training transformers for different systems and ensures accurate predictions without compromising efficiency. To validate the framework's effectiveness, we conducted a case study on screening amine-based solvents for carbon capture. This case study demonstrated the framework's efficiency in predicting properties under varying conditions and enabling engineers to perform solvent-screening tasks specific to the nature of the carbon capture process.

References:
[1] Sophie Sitter, Qi Chen, and Ignacio E. Grossmann. 'An Overview of Process Intensification Methods.' Current Opinion in Chemical Engineering, vol. 25, 2019, pp. 87-94.
[2] Guzman-Urbina, Alexander, et al. 'Systematic Process Energy Optimization via Multi-level Heat Integration: A Case Study on Low-Temperature Reforming for Methanol Synthesis.' Computer Aided Chemical Engineering, edited by Yoshiyuki Yamashita and Manabu Kano, vol. 49, Elsevier, 2022, pp. 1207-1212.
[3] Chen, Guangying, et al. 'Artificial Neural Network Models for the Prediction of CO2 Solubility in Aqueous Amine Solutions.' International Journal of Greenhouse Gas Control, vol. 39, 2015, pp. 174-184.
[4] Haghighatlari, Mojtaba, et al. 'Learning to Make Chemical Predictions: The Interplay of Feature Representation, Data, and Machine Learning Methods.' Chem, vol. 6, no. 7, 2020, pp.

169ct Diversity-Driven, Bayesian Optimization of MOF Designs to Optimize Performance for Environmental Applications Involving NH_3 Adsorption

Tsung Wei Liu, Quan Nguyen, Adji Bousso Dieng, Diego Gomez Gualdron

Metal-organic frameworks (MOFs) are promising materials to engender technology-enabling properties in numerous applications. However, one significant challenge in MOF development is their overwhelmingly large design space, which can be intractable to exhaustively explore experimentally (and even computationally) to find optimal designs. Assessed by its ability to optimize three NH3 sorption-based performance metrics that depend on different extent on material chemistry and structure, herein we demonstrate a diversity-driven optimization framework (Vendi Bayesian optimization, VBO) as an efficient way to find diverse MOF designs that optimize chemistry and structure-dependent properties. An advantage of this diversity-driven approach is preventing committing to a single MOF design (or similar ones) that for one reason or another may not be experimentally realized and/or tested. With ten simulated campaigns done for each case, we first statistically demonstrated that our VBO consistently does equal or better than random search to find high-performing designs within a 1,000-MOF subset for *i)* ammonia storage, *ii)* ammonia removal from membrane plasma reactors, and *iii)* ammonia capture from air. Then, with one campaign dedicated to ammonia storage on a "hybrid" 10,000-MOF database, we identify twelve extant and eight hypothesized designs that optimize ammonia working capacity ΔN_{NH3} between 300 K and 400 K at 1 bar. The best MOF designs are predicted to achieve ΔN_{NH3} values between 23.6 and 29.3 mmol/gm, potentially surpassing those that MOFs previously experimentally tested for NH_3 adsorption would have at the proposed operation conditions. The best hypothesized MOFs were predicted to be likely synthesizable (via free energy calculations), whereas the best MOF designs in general are predicted to be thermally stable (via machine learning prediction) and to require only ca. 10% of the energy content in NH_3 to release the stored molecule from the MOF. Finally, the analysis of the top MOFs indicates a pore size of around 10 Å, a heat of adsorption around 33 kJ/mol, and the presence of Ca as MOF design rules that could help optimize ammonia working capacity at the proposed operation conditions.

169cu Leveraging Grand Canonical Monte Carlo and Machine Learning for Classification of Solvent Affinity in Functionalized Polymer Membrane Materials

Andres Ordorica, Co Quach, Zane Parkerson, G. Kane Jennings, Clare McCabe, Peter Cummings

Understanding the affinity of polymers towards different solvents is crucial for the development of polymeric membranes for separation processes. In this study, we utilized Grand Canonical Monte Carlo (GCMC) simulations to assess the absorption behavior of water and ethanol on a target polymer. Leveraging the scripting capabilities of the MosDeF software suite, an extensive screening of functionalized dicyclopentadiene norbornene diacyl chloride (NBDAC) monomers was conducted. From the GCMC simulation results, the relative affinity of each functionalized NBDAC monomer towards water/ethanol can be quantified and the data used to train a classification machine learning model that can predict the polymers' preference for each solvent species. The model is trained on molecular descriptors derived from the cheminformatics library RDKits and simulation system details, capturing intricate intermolecular interactions governing solvent adsorption on the functionalized polymer membrane material. Through rigorous validation and evaluation, the efficacy of the machine learning approach in accurately categorizing polymer materials based on their solvent preference is demonstrated. This interdisciplinary methodology bridges the gap between molecular simulation techniques and data-driven approaches, providing insights into the solvent affinity of functionalized polymer membrane materials with implications for membrane design and separation processes. Our findings underscore the potential of combining computational simulations with machine learning for efficient characterization of membrane material properties.

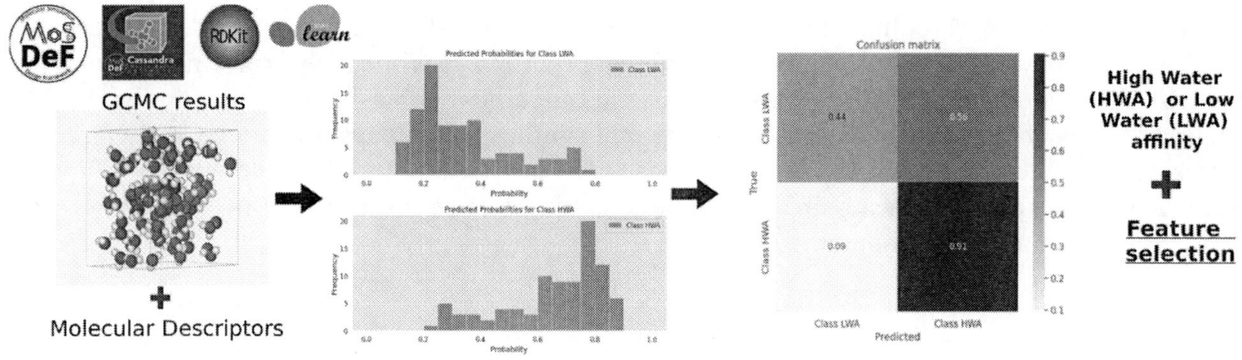

169cv Expanding Chemical Synthesis Planning to Explore Chemo-Enzymatic Pathways Using Minimal Transitions

Mohit Anand, Vikas Upadhyay, Costas D. Maranas

Chemo-enzymatic pathway design merges complimentary enzymatic and chemical reactions strengths to expand the biomolecular design space. While chemical reactions struggle with regioselectivity and stereoselectivity, biological processes encounter limitations such as product toxicity and enzyme scarcity. Optimally integrating both approaches provides an opportunity to identify efficient pathways beyond the capabilities of either approach. Recently, numerous studies have shown the advantages of leveraging enzymatic steps into industrial-scale chemical processes such as for the blood sugar regulator Sitagliptin (Merck) and for the treatment of AIDS, a protease inhibitor Darunavir (Prozomix). Designing optimal chemo-enzymatic pathways is a complex task and requires navigating a high-dimensional search space of potential reactions that combines individual chemical and biochemical steps. Unlike retrosynthesis which recursively traces back from a given target molecule to the starting precursor one step at a time, we solve a mixed-integer linear programming algorithm (MILP) to discover entire pathways by traversing reaction rules of known chemical and enzymatic connections extracted from USPTO and MetaNetX databases respectively. The algorithm minimizes the number of transitions between chemical and biological reactions, hence cutting down on multiple stages of separation and purification required for each switch between enzymatic and chemical reaction environments. Our approach has been validated through case studies on four molecules: 2-5-Furandicarboxylic acid, Terephthalate, 3-Hydroxybutyrate, and α-Pinene, which demonstrates our capability to identify previously reported total synthesis pathways while also exploring alternative routes with different chemical reaction conditions and outcomes such as temperature and yield respectively. This effort contributes to search for total synthesis pathways leveraging known reactions and streamlines the planning of chemical synthesis potentially leading to the development of more effective synthesis strategies, accessible through open-source code for users. (https://github.com/maranasgroup/chemo-enz).

169cw Integrating Machine Learning with Evolutionary Algorithms to Design and Discover High-Performing MOFs for Methane Adsorption Using Building Blocks and Crystal Information

Nicole Beauregard, Ranjan Srivastava

Metal-Organic Frameworks (MOFs) are an alluring class of porous crystalline materials. These materials are built from repeated coordinated units of metal clusters and organic linkers. By intentionally selecting what molecular building blocks to use to fabricate these materials, certain properties can be tuned for certain applications. One particular application of interest is gas adsorption. MOFs can be fabricated with high surface area and porosity as well as tunable pore size and functionality which are all desirable for gas adsorption. Methane is of specific interest as it is the main component of natural gas. Adsorbed natural gas (ANG) is a safer and more energy efficient storage method than conventional compressed natural gas (CNG) or liquified natural gas (ANG). Natural gas has been touted as a transition fuel from gas powered to electric vehicles yet these traditional storage techniques are not economically feasible for vehicular applications. Identifying high performing methane adsorbents would greatly reduce this barrier. The vast number of molecular building blocks available also results in a near limitless number of MOFs that could theoretically be fabricated. There could very likely be an unrealized MOF that exceeds current methane adsorption performance.

The goal of this work was to construct and screen high-performing metal-organic frameworks for methane storage. This feat was accomplished through the development of an integrated genetic algorithm (GA) and random forest (RF) machine learning algorithm (GARF). GARF is able to build hypothetical MOFs and predict their methane adsorption using only their molecular building blocks and crystal structure information. While it is infeasible to experimentally fabricate the near limitless number of possible MOFs, MOFs can be constructed and evaluated in silico. By using GARF to specify the building blocks and crystal information, a hypothetical MOF crystal can be generated. Comprehensive molecular simulations can then be employed to calculate methane adsorption. While faster than experiments, simulations are hindered by long convergence times and large computational cost. Machine learning (ML) offers a faster method for predicting methane adsorption. The downside with ML is that there is often a compromise with accuracy. Pardakhti et al. discovered that using structural and chemical properties as descriptors for a RF machine learning algorithm led to highly accurate methane adsorption predictions (1). In addition to accurately predicting adsorption, high-throughput screening of MOFs requires an intelligent method for generating MOF crystals. Evolutionary algorithms can be used to evolve high-performing solutions to problems using the principles of recombination and mutation. GAs, a class of evolutionary algorithms, encode solutions in a data structure known as a chromosome. These chromosomes are allowed to evolve until a solution meeting the desired criteria is found or some other termination criteria is met. The chromosome consists of the building blocks and crystal information needed to build a MOF crystal. In order to function efficiently, GAs require an efficient accurate fitness function to evaluate the chromosomes. In this case, the GARF algorithm was used to replace comprehensive molecular simulations with RF machine learning. GARF was able to screen 250,000 MOFs in mere minutes on a personal computer significantly reducing the computational time while also effectively screening these hypothetical MOFs for their methane adsorption. The resulting decrease in both time and computational cost would be invaluable for materials discovery.

Using solely molecular building blocks and crystal information, the GARF algorithm was able to successfully evolve high-performing MOFs for methane adsorption. Figure 1 illustrates the progression from building blocks to methane adsorption. First, GARF uses the building blocks and crystal information which includes framework number and topology to generate a hypothetical MOF crystal. From this crystal, a chemical formula can be extracted which can be used to calculate several chemical properties. The crystal information and chemical information are then used as descriptors to predict six structural properties. These structural properties along with the chemical properties are used as descriptors to predict methane adsorption. To train the methane adsorption and six structural property random forest algorithms,

50,000 hypothetical MOFs (hMOFs) from the MOFXDB database were used (2). This database contains values for structural properties and methane adsorption calculated via comprehensive molecular simulations along with crystallographic information files (CIFs) detailing their crystal structure. Using this training data, we achieved R2 values above 0.92 for the six structural properties and methane adsorption and relatively low mean absolute error percentage (MAPE) values justifying the use of the RF algorithms for as the fitness evaluator for the genetic algorithm. The GARF model was then able to evolve high-performing MOFs with chromosomes matching MOFs in the top 100 of the hMOF database. After extracting an ensemble of high-performing evolved MOFs generated by GARF, we were able to elucidate information about the ideal chemistries and building blocks for generating MOFs with high methane adsorption. This will allow us to intelligently select new building blocks to add to GARF to evolve novel MOFs. The GARF algorithm rapidly and accurately screens MOFs using minimal input information and evolves high-performing candidate MOFs for methane adsorption opening up a new avenue for materials discovery.

REFERENCES

1. Pardakhti, M., Moharreri, E., Wanik, D., Suib, S. L. & Srivastava, R. Machine Learning Using Combined Structural and Chemical Descriptors for Prediction of Methane Adsorption Performance of Metal Organic Frameworks (MOFs). ACS Comb. Sci 19, 34 (2017).
2. Bobbitt, N. S. *et al.* MOFX-DB: An Online Database of Computational Adsorption Data for Nanoporous Materials. *J. Chem. Eng. Data* **68**, 483–498 (2022).

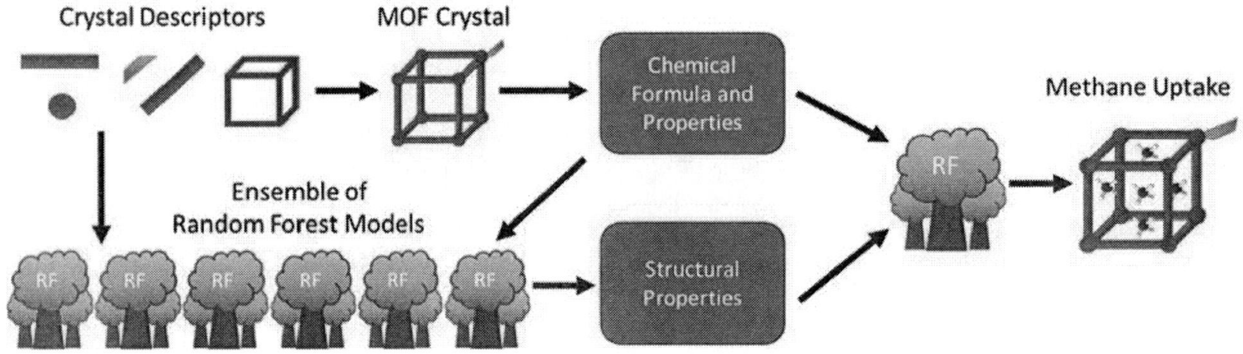

169cx Developing an Open-Source Tool for Generating Rich and Consistent Sigma Profiles

Fathya Salih, Dinis Abranches, Yamil Colón

The field of computational material discovery using machine learning has seen a lot of development in terms of novel neural network architectures, advanced training protocols, and novel molecular descriptors. However, the latest of these has an unsolved issue. All the commonly used molecular descriptors (SMILES strings, graphs, positional coordinates, etc.) share the problem of being dependent on molecule size. However, for most deep learning models, the size of the input has to be fixed to the size of the largest molecule in the dataset. Meaning, the network will end up with many unnecessary parameters and will require a larger training dataset. Additionally, the neural network that may have taken months to train will have an upper limit for the size of molecule it can be used with in the future.

As an alternative, sigma profiles have been proposed as a universal molecular descriptor. A sigma profile is an un-normalized histogram of the surface screening charge distribution of a molecule when embedded in a continuum solvent with a very high dielectric constant. Therefore, sigma profiles contain valuable information about the possible inter- and intra-molecular interactions the molecule can experience. And they have been successfully used as inputs to convolutional neural networks and gaussian processes to predict bulk material properties (e.g. boiling point, density, aqueous solubility, etc.) with minimal datasets of less than 1500 molecules. Furthermore, because they are histograms, their size is independent of molecule size - the sigma profiles for methane and C30 will both have the same number of bins. The main reason sigma profiles have not been used as molecular descriptors in more machine learning applications so far is the lack of an open-source tool for generating them, and the publication limits placed by commercially available tools.

This work summarizes the development of an open-source python tool for generating sigma profiles to allow for their use in large-scale materials discovery. The work, also, addresses the effect of conformers on sigma profiles to ensure the generated profiles are consistent and rich with information. The quality or 'richness' of the sigma profiles produced was benchmarked against datasets from literature by comparing the performances of Gaussian process models at predicting material properties with the different sigma profile sources.

169cy Unsupervised Computational Analysis of Major Histamine Complex II-Binding Epitope Sequences

Hannah Hargrove, X. Frank Zhang

Abstract

The major histamine complex II (MHC-II) receptor class is critical to immune reaction in the human body [1]. A lack of MHC-II activation has been linked to increased cancer immune evasion [2], and many studies in recent years have focused on the relationship between functionality of MHC-II receptors and decreased functionality of many varieties of tumors [3] [4] [5] [6]. Many therapeutic drugs designed to influence immune reaction target MHC-II receptors. By analyzing epitope sequences of proteins known to bind to MHC-II receptors, a more thorough understanding of key sequence fragments for MHC-II receptor binding can be obtained. With a more thorough understanding of which sequence fragments are critical to binding, a more effective MHC-II-targeting drug candidate can be designed *in silico* faster than traditional *in vitro* methods, at a lower cost.

1. **Introduction**

Adaptive immunity refers to the host immune system's ability to adapt to novel threats over time, and major histamine complex (MHC) receptors are critical to this process [7]. There are two broad subclasses of MHC: MHC class I (MHC-I) and MHC class II (MHC-II). Both of these subclasses are glycoproteins, and they present antigenic peptides to T cells for destruction. MHC-I presents antigenic peptides to CD4+T cells, and MHC-II presents them to CD8+T cells. The major difference between MHC-I and MHC-II receptors is MHC-I glycoproteins are expressed on the surface of nearly all nucleated cells, whereas MHC-II glycoproteins are expressed primarily on the surface of antigen presenting cells (APCs) [8]. Due to slight differences in functionality between MHC-I and MHC-II, this study only focuses on MHC-II, to avoid excessive generalization between subclasses of MHC.

MHC-II molecules are produced in MHC class II compartments (MIICs), which form many different subclasses of MHC-II and dictate each molecule's function [9]. Overall, MHC-II molecules are membrane anchored, with MHC-encoded, glycosylated alpha and beta subunits. They usually bind peptides 10-15 amino acids in length, but can bind much longer peptides, with hundreds of amino acids [7]. The MHC class of molecules is defined by its polymorphic nature – each individual MHC molecule could present hundreds if not thousands of different peptide antigens. The extent of this polymorphism varies across different vertebrate species, but is present in all vertebrates, and the total number of molecularly defined allelic variants for human MHC-I and MHC-II is in the thousands [7].

The generalized structure of MHC-II is well-understood, including features of peptide binding pockets. In general, MHC-II has alpha and beta subunits which come together to present antigenic peptides for targeting by T cells. Figure 1 depicts a ribbon model of the MHC-II molecule, supporting a model antigenic peptide depicted in a ball-and-stick model form. The two subunits join to form a cup-like structure, which allows the antigenic peptide to be presented outside the cell for T cell action [7].

Although MHC-II molecules can be grouped broadly as a specific class of molecules, there are important subclass distinctions which need to be considered when analyzing MHC-II-antigen binding. Additionally, each subclass performs a slightly different MHC function [9]. This is due to the highly polymorphic and varied characteristics of MHC molecules as a whole. To account for this, this study will impose limits on what molecules will be considered. Firstly, the study will be limited to MHC-II only. This will allow the behavior of MHC-II molecules to be analyzed as a broad category. Although MHC-II molecules are incredibly diverse, there may still be some generic patterns observed. After generalized assessment of MHC-II molecules as a class, analysis of several subtypes of MHC-II will be conducted to determine the differences between disparate types of MHC-II/antigen binding.

When splitting the MHC-II class into subcategories, it is important to consider the limitations of the database in use. This study relies upon the Immune Epitope Database (IEDB) [10], which is capable of splitting assay results by several different experimental factors. Only linear epitopes were considered, as opposed to conformational epitopes, as this eliminates the need for structural data. To further narrow the study, data was limited to human leukocyte antigen (HLA), a specific subclass of MHC-II. Finally, the data was split into positive and negative sets, positive indicating MHC-antigen binding took place and negative meaning no significant binding took place. This was done so unsupervised machine learning (UML) could be used to analyze the data for biases and pattern recognition.

After data was obtained from the IEDB, further preprocessing was conducted. Any epitopes which had additives branching off the main epitope sequence were removed. Initial analysis of the dataset revealed an overabundance of polyamino acids, so any epitopes which had more than 4 duplicate acids in a row were removed to avoid skewing the model. With this, the data is ready to undergo unsupervised assessment.

The model used in this study is known as N-gram analysis, which is a subclass of UML. This analysis type relies on tokenization of a string of data, which is a method of breaking down a long string of data into manageable chunks, which can then be analyzed by the unsupervised learning model. By tokenizing the antigen epitope sequences, the unsupervised learning model can then conduct statistical analysis of the dataset as a whole, and determine any emergent patterns and trends over thousands of instances. In the context of UML, a single data point in a dataset is referred to as an **instance**, and the variables which distinguish various instances from each other are called **attributes**.

This study primarily focused on unsupervised analysis of positive binding epitopes. Future work is to include analysis of negative binding, and using the combination of both to construct a supervised model capable of predicting HLA-binding of a novel sequence.

2. Results

The first stage of analysis focused on observing the generic HLA-MHC-II binding epitope dataset. This dataset underwent unsupervised N-gram analysis to determine the most common N-grams for generic HLA epitopes. Values of N between 1 and 8 were analyzed. From this analysis, a vocabulary graph can be constructed, as depicted in Figure 2. As can be observed from this graph, lower values of N result in a vocabulary plateau. This makes sense, because due to the method of tokenization, there are a limited number of novel tokens which could possibly exist – for example, since there are 20 possible amino acids, there are only 20 possible 1-grams, and 400 possible 2-grams, and so on. As values of N get higher, the plateauing behavior is seen less prominently, which indicates the dataset is not diverse enough to encompass the full scope of possible 8-grams, for example (20^8 possible tokens).

Based on Figure 2, N-grams of N = [1,4] result in a saturated vocabulary. Based on this it can be assumed that the sample size over that range of N is of adequate size for analysis of the dataset at that value of N. For larger values of N, the vocabulary does not reach a saturation plateau before termination of analysis, and therefore it is more likely that novel instances will appear. In light of this, the most significant N-gram sequence predictions of epitope binding will likely come from the range of N = [1,4].

After analysis of the generic HLA-binding epitope dataset, the set was split into disparate subtypes of HLA, which included HLA-DQ, HLA-DR, and HLA-DP. Each of these sets was subsequently analyzed to determine the most likely N-Grams in each set, with values of N = [1,8]. As seen in Figure 3, both HLA-DR and -DP exhibit similar populations of unigrams (1-grams) as the generalized HLA dataset, with the four acids L, A, K and E appearing in the top six most common acids for each subtype. However, HLA-DQ exhibits a drastically different breakdown of common amino acids, with the top six most common acids being A, E, G, S, P and V. This would seem to imply different properties are necessary to bind with HLA-DQ than with other HLA types.

Additional analysis on larger values of N-gram tokens are ongoing for each of these datasets, but from a preliminary standpoint there seem to be certain tokens which appear a statistically significant amount more than random chance would dictate.

3. Conclusions and Future Work

The most common amino acids in molecules which bind to HLA receptors in general are L, A, K, and E, and it is evident certain types of amino acid are favored for HLA-binding over others. This is an important first step towards allowing the initial stage of MHC-II-targeting drug design to be performed in-silico rather than in-vitro, which could save time and money.

The determination of which single amino acids are most common in HLA-binding proteins is an important first step towards developing an in-silico prediction model for HLA-binding. Future work is to include constructing a natural language processing model which takes in a user-provided amino acid and uses the HLA datasets to predict the configuration of acids most likely to bind to an HLA receptor. In-vitro experimentation will be conducted to verify these results, and to further improve the model's predictive capabilities.

Figure 1

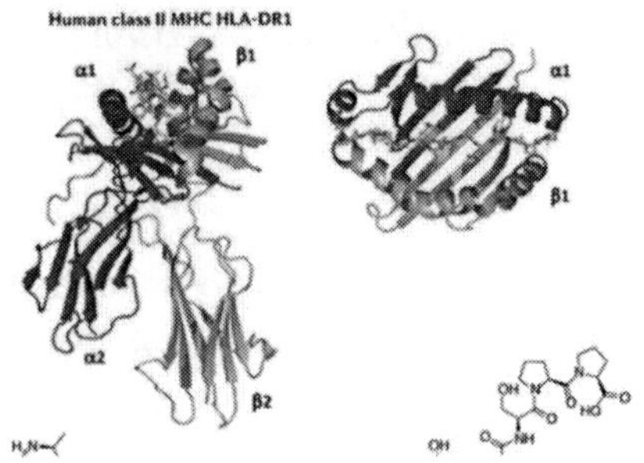

Figure 1: 3-dimensional ribbon model structure of the MHC-II molecule, with a model antigenic peptide presented in ball-and-stick model form. Figure produced by N. Pishesha et. al., 2022 [illegible].

Figure 2

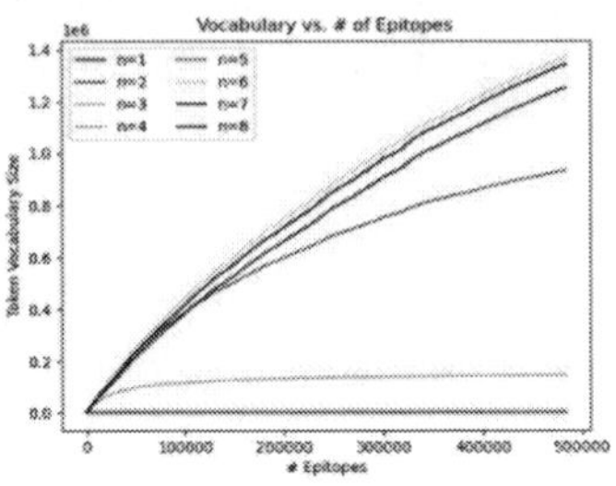

Figure 2: Vocabulary graph of N-gram analysis for values of N 1-8. The dataset under consideration is the generic HLA epitope dataset.

Figure 3

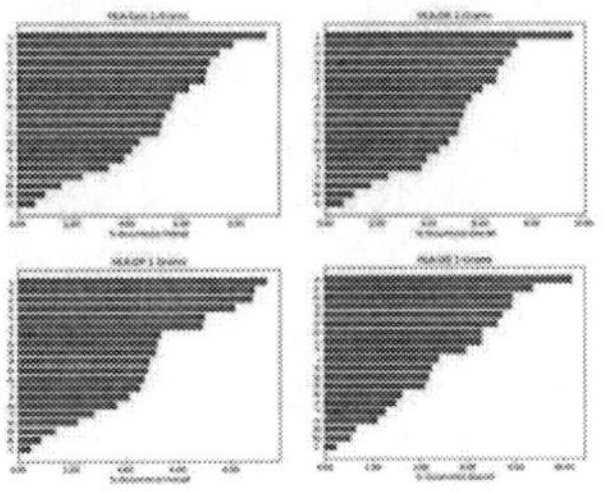

Figure 3: Four bar charts depicting the number of times each amino acid appeared in the dataset, as a percentage of the total number of amino acids.

169cz AI Based Exploration on Synthesizable Space for Autonomous Laboratory

Nayeon Kim, Hyuk Jun Yoo, Donghun Kim, Sang Soo Han

Autonomy laboratories have been recently conducted widely for accelerating material development. In traditional research, humans directly define the initial experimental conditions, which are constraints for solving such optimization problems, thereby slowing down the development of materials. Furthermore, under the initial experimental conditions as defined synthesizable space found by humans, it may not be possible to find the global optimal. Therefore, we solve the problem of requiring a human to set the initial experimental conditions by developing a model that breaks the hardware constraints and explores the domain using active learning. Our research demonstrated that superior performance compared to other baseline models in an unexplored synthesizable area before conducting experiments, enabling efficient exploration in fewer experiment iterations. This study proves the justification of applying exploration on synthesizable space which can applicability to new environments and experimental conditions, so that our work verified the conventional autonomous lab even more flexible.

169da Predicting Ionic Conductivity of Ionic Liquid and Solvent Mixtures Using Machine Learning

Masrur Ahmed, Jindal Shah

Ionic liquids (ILs) are a special type of molten salts that are molecular in nature and therefore can be designed to exist as liquids at low temperatures. Their high thermal and electrochemical stability, negligible volatility, low melting point and the capability to be tuned for various target properties make them advantageous to be used as solvents in industrial processes. However, low ionic conductivity, high viscosity, and high market price of many ionic liquids are a hindrance to their widespread usage. One of the ways these limitations can be addressed is by mixing ionic liquid with an organic solvent. Although there exists some work with specific IL-solvent combinations, the lack of a general predictive model encompassing various solvents and ionic liquids implies that a large number of models need to be developed for such combinations. In this work, we showcase our attempt at developing a unified ionic liquid-solvent machine learning model to predict ionic conductivity of ionic liquid-solvent mixtures. Specifically, we apply three machine learning models Random Forest, XGBoost, and Graph Neural Network coupled with molecular and graph-based structural features. We assess the effect of featurization techniques and identify important features governing ionic conductivity of ionic liquid-solvent mixtures. We will also show how these models enable high-throughput screen to identify ionic liquid-solvent mixtures exhibiting high ionic conductivity.

169dc Representation of Stochastic Polymeric Materials for Machine Learning Applications

Ludwig Schneider

Here is a revised version of your abstract:

Polymeric materials are versatile, with applications ranging from nano-science to everyday items. For industrial purposes, these materials often exhibit significant stochastic variations in molecular weight and composition. Representing such materials for machine learning purposes and computational studies remains a challenge. In this presentation, we explore a stochastic graph-based approach that can be constructed from BigSMILES and G-BigSMILES string representations. These graphs are then encoded via graph autoencoders to represent them in a continuous R^N space. This representation can serve as an initial point for further machine learning applications, such as deep neural nets, databank searches, or can be interpolated for new materials with generative AI.

169dd Cross-Database Discovery of Metal-Organic Frameworks with Open Cu(II) Sites for Biogas Upgrading through Machine Learning

Xiaoyu Wu, Jianwen Jiang

This study presents a systematic approach to screen Metal-Organic Frameworks (MOFs) with open Cu(II) sites (OCS) for biogas upgrading, by integrating machine learning (ML) with a novel validation strategy. Distinguishing itself in both methodology and application scale, our work employs force fields tailored for the specific interactions between OCS and CO_2/CH_4 binary-component system, enhancing screening accuracy for biogas separation using unique OCS-MOFs meticulously sourced and curated from the Ab-initio REPEAT Charge (ARC) MOF database.

Our work further delves into the application of ML to this field. By tracing the origins of MOFs within the ARC-MOF database, which amalgamates various sub-databases, into focused subsets of OCS-MOFs, we not only broaden the scope of potential MOF candidates but also introduce a rigorous external validation step for the integration of data-science heuristics in materials discovery. This involves testing the ML model against a set of databases not used in conventional ML training/test sets, offering a stringent assessment of its predictive capability and generalizability across unseen chemical structures.

The methodological novelty of our work lies in this dual layer of innovation: the employment of OCS-specific force fields for more accurate description of interactions and the systematic external validation of ML predictions. Together, these advancements contribute significantly to the fields of materials science and chemical engineering, offering a reliable and nuanced approach to the discovery and evaluation of MOFs for environmental applications, particularly in the context of biogas upgrading.

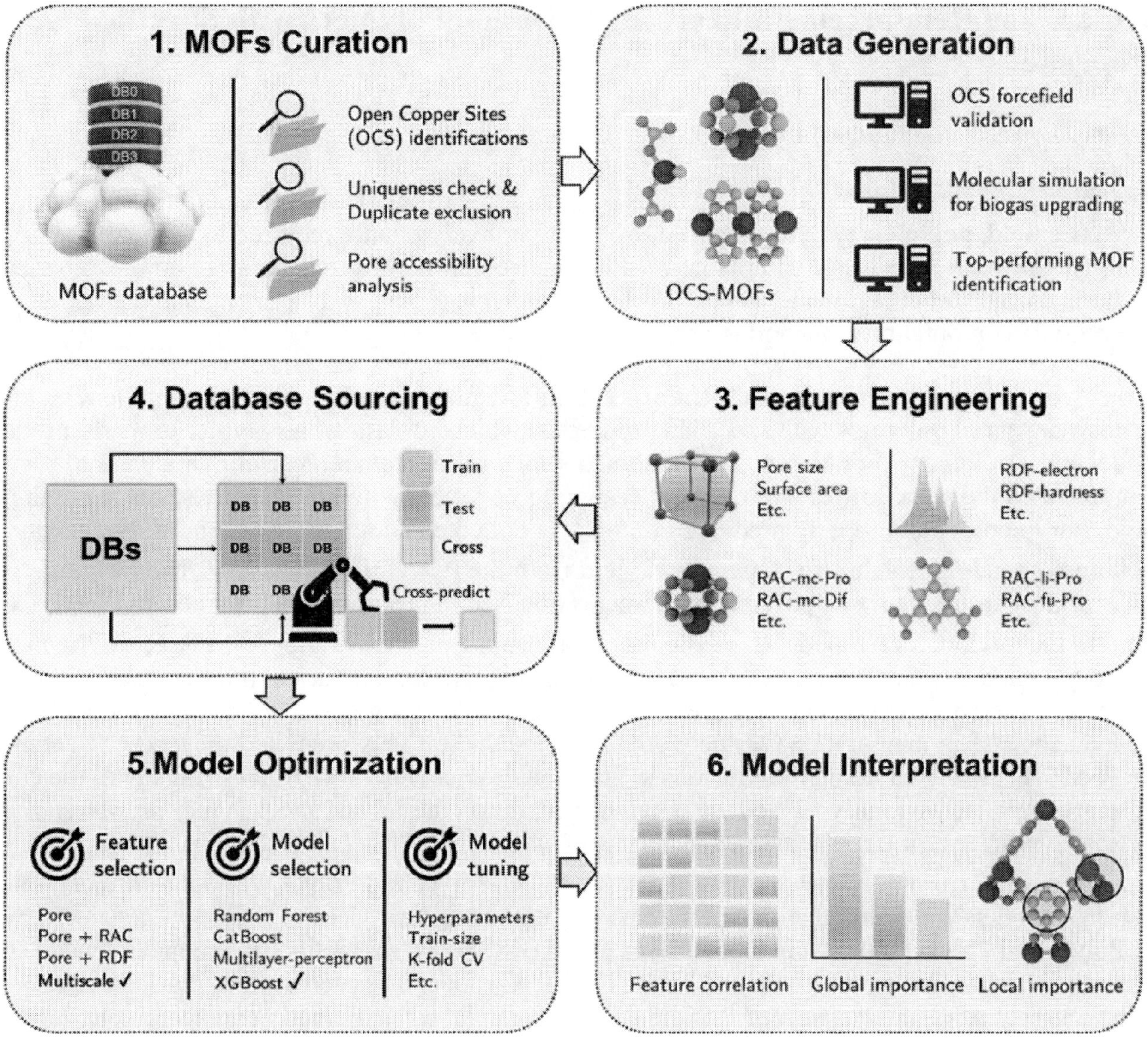
1. MOFs Curation
DB0
DB1
DB2
DB3
MOFs database
Open Copper Sites (OCS) identifications
Uniqueness check & Duplicate exclusion
Pore accessibility analysis
2. Data Generation
OCS-MOFs
OCS forcefield validation
Molecular simulation for biogas upgrading
Top-performing MOF identification
3. Feature Engineering
Pore size
Surface area
Etc.
RDF-electron
RDF-hardness
Etc.
RAC-mc-Pro
RAC-mc-Dif
Etc.
RAC-li-Pro
RAC-fu-Pro
Etc.
4. Database Sourcing
DBs
DB
Train
Test
Cross
Cross-predict
5.Model Optimization
Feature selection
Pore
Pore + RAC
Pore + RDF
Multiscale ✓
Model selection
Random Forest
CatBoost
Multilayer-perceptron
XGBoost ✓
Model tuning
Hyperparameters
Train-size
K-fold CV
Etc.
6. Model Interpretation
Feature correlation
Global importance
Local importance

169de Using Reinforcement Learning to Design Polymers with Specified Properties

Haifan Zhou Sr., Yue Fang, Hanyu Gao

Designing polymers with specified properties is crucial in various fields, for example, in the aerospace field, polymers with high tensile stress strength at yield are required to maintain device stability. Currently, the design of polymers with specified properties relies mainly on the experience of chemists and wet experiments, which is often a time-consuming and labor-intensive process. Therefore, a computational method is needed to improve efficiency.

In this work, we constructed a model RLPolyG (Reinforcement Learning for Polymer Generation) for de novo design of polymers with specified properties, which consists of a polymer property prediction model and a polymer generation model combined with a reinforcement learning model. We demonstrated the workflow of RLPolyG by designing polymers with high tensile stress strength at yield. For the polymer property prediction model, we collected a total of 267 homo-polymers and their corresponding tensile stress strength at yield from the PoLyInfo database[1]. The Morgan fingerprint with frequency of polymer monomers was used to represent the polymer and served as input to the random forest model to predict its tensile stress strength at yield[2]. The R^2 of the model on the train and test dataset were 0.92 and 0.94, respectively, demonstrating the accuracy of the model. For the polymer generation model, we constructed an autoregressive language model based on the long short-term memory(LSTM) network. The model iteratively predicted the next character in the PSMILES string based on the previous k characters to generate PSMILES strings with the correct structure. Finally, we jointly trained the polymer generation model and the polymer property prediction model with reinforcement learning to generate polymers with specified properties[3]. We compared the performance of polymers generated by RLPolyG and PolyG(without reinforcement learning) models and found that the mean tensile stress strength at yield of polymers generated by RLPolyG and PolyG model were 0.05783Gpa and 0.04888Gpa, respectively. The mean tensile stress strength at yield of polymers generated by the RLPolyG model was increased by 18.31%. The computational results demonstrated that the RLPolyG model can efficiently and accurately design and optimize novel polymers with specified properties.

Reference:

[1] Otsuka S, Kuwajima I, Hosoya J, et al. PoLyInfo: Polymer database for polymeric materials design[C]//2011 International Conference on Emerging Intelligent Data and Web Technologies. IEEE, 2011: 22-29.

[2] Yue T, He J, Tao L, et al. High-throughput screening and prediction of high modulus of resilience polymers using explainable machine learning[J]. Journal of Chemical Theory and Computation, 2023, 19(14): 4641-4653.

[3] Popova M, Isayev O, Tropsha A. Deep reinforcement learning for de novo drug design[J]. Science advances, 2018, 4(7): eaap7885.

169df Efficiently Screening Metal-Organic Frameworks Via Molecular Simulation with Multi-Armed Bandit Algorithms

Qia Ke, Jana Doppa, N. Scott Bobbitt, Cory Simon

Metal-organic frameworks (MOF) are nanoporous materials composed of metal ions/clusters and organic ligands. They have large specific surface areas and a high density of functional groups. These features allow MOFs to store, sense, separate, and controlled-release gases. The porosity, pore size and shape, and chemistry of MOFs are highly adjustable. Among millions of conceivable MOFs for a given adsorption task, we can screen them using molecular simulations. However, brute-force screening is often infeasible in terms of cost and time.

This work applies multi-armed bandit optimization (MABO) to leverage short, cheap, and noisy molecular simulations to screen a large set of MOFs while incurring the least computational cost. Each MOF is analogous to a slot machine; conducting a Monte Carlo molecular simulation is analogous to pulling an arm of a slot machine; and the noisy, estimated property of a MOF from a simulation is analogous to receiving a stochastic reward from a slot machine. MABO adaptively selects a sequence of MOFs for cheap, noisy simulations while learning from feedback and balancing exploration and exploitation in its decision-making. After each simulation, MABO uses the noisy estimate of this MOF's property to update a posterior distribution of that MOF's property. Then, MABO selects the next MOF for a simulation based on the posterior distributions of the property of all candidate MOFs. Its decision balances the yearning to select a MOF with (i) a high mean property (exploitation) and (ii) a high variance (i.e., uncertainty) in the property (exploration).

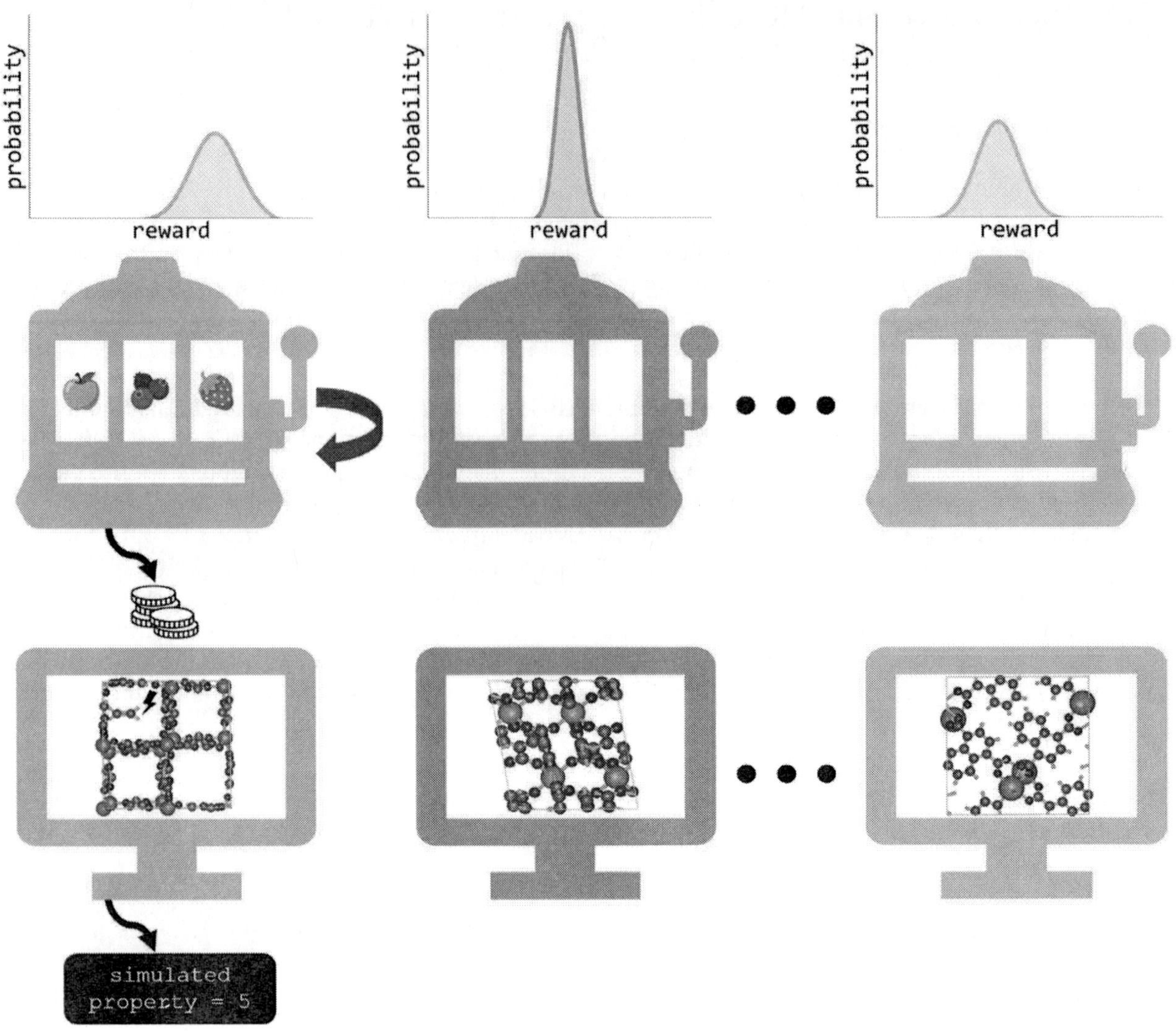
probability
reward
probability
reward
probability
reward
simulated
property = 5

169di Using Enhanced Sampling Methods to Elucidate the Mechanism of Noncanonical Redox Cofactor Dependent Engineered Enzymes

Parth Bandivadekar, Surl-Hee Ahn, Han Li, Yi Ping

Background: Enzymes are useful proteins that can also produce compounds renewably and under ambient conditions while generating low waste. However, they often require natural cofactors or small, non-protein molecules for optimal activity. These natural cofactors are prohibitively expensive and chemically unstable to be used in specialty chemicals manufacture [1]. Hence, researchers have been engineering enzymes that can utilize smaller natural cofactor mimetics or noncanonical cofactors that are more efficient, easier to synthesize, and are low-cost alternatives.[2]

Dr. Han Li (UC Irvine) recently engineered the enzyme phosphite dehydrogenase (PTDH) to utilize a noncanonical cofactor 1-benzylnicotinamide (BNA+) in biotransformation processes [3]. Dr. Li also produced PTDH mutants that exhibit higher activity compared to the wild type engineered PTDH. However, the reason for that is unknown, and it has been hypothesized that these mutations are space filling that pack against each other where the natural cofactor would have bound, which leads the PTDH mutants to mimic the binding of a natural cofactor and become more active closed proteins. To harness the PTDH mutants fully at industrial scale, we must understand the molecular details and mechanisms of how the mutations lead the enzymes to become more active closed proteins.

Aim: 1) MD Simulations of two PTDH mutants and the native PTDH to investigate the higher activity caused by additional mutations in LY-13-Mut than LY-13.
2) MD Simulations of two PTDH mutants and the native PTDH transitioning from the initial open state to the bound closed state to uncover the closing mechanisms and rates.

Methodology: We outlined the simulations of the two highly active PTDH mutants LY-13 and LY-13-Mut and the wild type PTDH 4EBF, which will provide detailed views into the functional mechanisms of PTDH. Specifically, we used 1 μs long Gaussian accelerated molecular dynamics (GaMD) that adds a harmonic boost potential to the system so that the energy barriers between states are reduced and the energy landscape can be sampled efficiently [4].

MD simulations are run using 10^{-15} s time steps, due to being limited by the fastest motions in the system (e.g. vibration of bonds), while protein dynamics range from 10^{-6} s to longer. Thus there is a major timescale barrier between MD simulations and protein dynamics. To bridge the timescale barrier, "enhanced sampling methods" for MD simulations were used. Specifically, we used Gaussian accelerated molecular dynamics (GaMD) that adds a harmonic boost potential to the system so that the energy barriers between states are reduced and the energy landscape can be sampled efficiently [4].

Although GaMD is useful for obtaining thermodynamic properties like free energy landscapes, other approaches need to be used to obtain kinetic properties like rate constants and continuous pathways since GaMD alters the real kinetics of the system with its biasing potentials. Hence, we will use the weighted ensemble method (WE), a path sampling method [5] , to fully characterize the closing mechanisms and rates for the PTDH systems starting from their initial open states. Dr. Ahn has successfully applied WE to investigate the activation mechanism of the SARS-CoV-2 spike protein [6] and is confident to introduce and apply the method for systems important for chemical engineering.

Expected Results: We expect to be able to gain atomistic insight into each PTDH system's structure and dynamics. Specifically, we expect to identify how the two mutations of LY-13-Mut contribute to its higher activity than that of LY-13. We expect that LY-13-Mut will tend more towards the active closed state and/or keep BNA+ bound longer than LY-13, as experiments predict. Moreover, we expect to see the LY-13-Mut show a similar free energy landscape to that of the wild type PTDH, which would indicate that the mutations reshape the conformational landscape of LY-13-Mut to resemble that of the wild type PTDH, similar to the results from Dr. Li's previous work on the engineered NADH oxidase that can utilize noncanonical cofactors [5].

Implications: PTDHs are highly efficient catalysts in recycling noncanonical redox cofactors due to their superior TTN, ability to maintain turnover at low BNA+ concentrations, and capability to recycle other simpler cofactor biomimetics as established by Dr. Li [3]. However, elucidating the mechanism of redox cofactor dependent enzymes will further our understanding of engineering growth-based selected enzymes for biotechnology applications. This is especially crucial in engineering biomimetics for industrially important reactions that do not exist in nature. In this pursuit, enhanced sampling methods with MD simulations are paramount in helping develop a unified design toolkit for engineered enzymes.

REFERENCES

[1] T. Knaus, C. E. Paul, and et. al, Journal of the American Chemical Society 138, 1033 (2016).
[2] E. King, S. Maxel, and H. Li, (2020), https://doi.org/10.1016/j.copbio.2020.08.005.
[3] L. Zhang, E. King, W. B. Black, C. M. Heckmann, A. Wolder, Y. Cui, F. Nicklen, J. B. Siegel, R. Luo, C. E. Paul, and
H. Li, Nature Communications 13 (2022), 10.1038/s41467-022-32727-w.
[4] Y. Miao, V. A. Feher, and J. A. McCammon, Journal of Chemical Theory and Computation 11, 3584 (2015).
[5] G. Huber and S. Kim, Biophysics Journal (1996), 10.1016/S0006-3495(96)79552-8.
[6] T. Sztain and S.-H. Ahn, (2021), https://doi.org/10.1038/s41557-021-00758-3.

[7] E. King, S. Maxel, Y. Zhang, K. C. Kenney, Y. Cui, E. Luu, J. B. Siegel, G. A. Weiss, R. Luo, and H. Li, Nature
Communications 13 (2022), 10.1038/s41467-022-35021-x.

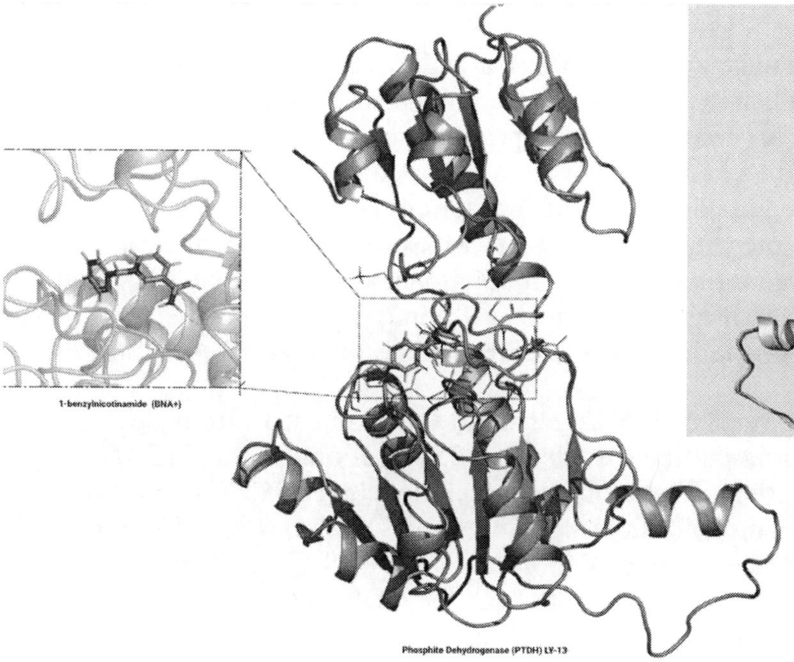

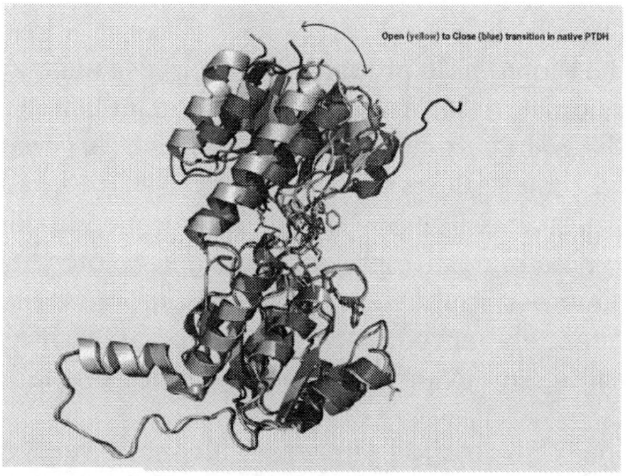

169dj Development of a Modifiable Atomistic Cellulose Nanocrystal Model

Arash Elahi, Santanu Chaudhuri

The remarkable properties of cellulose nanocrystals (CNCs), a widely used natural polymeric nanomaterial in various sustainable industries, result from the unique structural characteristics of CNCs. In particular, the trans-gauche (tg) conformation of primary alcohol groups on the cellulose molecules allows for the formation of an extensive network of intra-chain and inter-chain hydrogen bonds. The high population of hydrogen bonds within the interior domain of cellulose nanocrystals confers high stiffness and mechanical integrity, enabling their use as reinforcement material in numerous applications. However, most atomistic-scale computational studies of CNCs have failed to accurately reproduce the high population of tg conformations within the crystalline domain of CNCs, leading to deviations in the calculated crystalline properties from experimental observations.

The conformation of primary alcohol groups and hydrogen bond formation stem from a complex interplay of interactions between different particles, such as angle vibrations, torsional forces, electrostatic, and van der Waals interactions. Therefore, accurate modeling of CNCs demands the use of an appropriate force field and charge model. In the study presented herein, we used a combination of the OPLS-AA force field and CM5 charge model, denoted as the OPLS-CM5 model, and observed significant improvement in the atomistic-scale models of CNCs. The OPLS-CM5 model yielded a 90% occupancy of tg conformations and hydrogen bonds, enhancing the previous OPLS-AA model by nearly 50%.

Given the excellent capability of this model in maintaining the crystalline properties of CNCs, it could be extended to the simulation of surface-functionalized CNCs without affecting their crystalline properties—an issue that has hindered previous models. Understanding the dynamics of surface-functionalized CNCs enables tailoring their surface chemistry for specific applications.

169dk Temperature-transferable Coarse-grained model for Studying Pluronic Triblock Copolymers

Arash Elahi, Xavier Bidault, Santanu Chaudhuri

Pluronics are a class of triblock copolymers consisting of central hydrophobic propylene oxide (PO) blocks flanked by two hydrophilic ethylene oxide (EO) blocks. This amphiphilic nature makes Pluronics thermos-responsive, enabling their use in various industries, including wastewater treatment and drug delivery. However, the underlying dependence of thermo-responsive behaviors, such as critical micelle temperature (CMT), on the copolymers' composition or size remains unclear. Although atomistic-scale modeling is an appropriate approach to understand the underlying physics of the temperature-dependent micellization of Pluronics, its time and length scale limitations prevent it from capturing the complex micellization process, which typically occurs on the microsecond scale.

Thus, in this study, we adopted a coarse-grained molecular dynamics simulation approach using the MARTINI force field to understand the impact of temperature on the micellization of Pluronics, varying in size and composition. Since temperature-transferability is a widely known limitation of the MARTINI force field, we developed a temperature-dependent interaction between the Pluronic blocks and water blocks to overcome this limitation. The resulting coarse-grained model was able to capture the CMT of 6 Pluronics by forming micelles above CMT and remaining as unimers below CMT. The simulation findings were used to predict the CMT of various Pluronics based on their molecular weight and composition, successfully predicting the CMT of 11 Pluronic systems with a mean absolute error of 1.75°C.

169dl The Molecular Simulation Design Framework (MoSDeF): Enabling High-Throughput Simulations via Active Learning Integration Workflows

Nicholas C. Craven, Kieran Nehil-Puleo, Chris Jones, Co D. Quach, Eric Jankowski, Clare McCabe, Peter Cummings

The Molecular Simulation Design Framework (MoSDeF)[1] was developed as a suite of tools to mitigate challenges associated with initializing and conducting molecular simulations, with a focus on facilitating reproducibility.[2] By providing standardized, open-source libraries, MoSDeF can automate steps in the preparation of chemical/biological systems for simulation, thereby minimizing unnecessary human involvement and associated errors. Furthermore, the automation of workflows enables easier setup of high-throughput screening processes, through integration with workflow management tools such as signac and signac-flow.[3] To further leverage and extend these types of workflows, a new package, genGrouper, has been developed for generalized chemical representation of simulation inputs, utilizing the PyTorch deep learning library.[4] Additionally, progress has been made in supporting features necessary for more complex water models, such as TIP4P/2005f and TIP5P models, and a standard screening workflow centered around MoSDeF software has been performed and made available for facile extension to more sophisticated high-throughput active learning approaches.

References

1. "MoSDeF" [Online]. Available: https://mosdef.org
2. Cummings, P. T. *et al.* (2021). Open-source molecular modeling software in chemical engineering focusing on the Molecular Simulation Design Framework. In AIChE Journal (Vol. 67, Issue 3). Wiley.
3. Adorf, C. S., *et al.* Simple data and workflow management with the signac framework. Comput. Mater. Sci., 146(C):220–229, 2018.
4. Ansel, J., *et al.*, (2024) PyTorch 2: Faster Machine Learning Through Dynamic Python Bytecode Transformation and Graph Compilation. In ASPLOS '24: Proceedings of the 29th ACM International Conference on Architectural Support for Programming Languages and Operating Systems, Volume 2. ACM.

735b Bayesian Uncertainty Quantification of Transferable Coarse-Grained Models of Monosaccharides

Parisa Farzeen, Abhishek Sose, Soumil Joshi, Sanket Deshmukh

Carbohydrate-based materials are widely used in various applications due to their unique properties, such as solubility, biocompatibility, non-toxicity, biodegradability, and mechanical strength. To accelerate the design of novel carbohydrate-based glycomaterials with desired properties, the employment of accurate and robust all-atom and coarse-grained (CG) models used in molecular dynamics (MD) simulations is crucial. Specifically, ensuring the reliability of MD simulation predictions requires robust, True-Tightened force field (FF) parameters that not only consistently match physical predictions but even enhance our precision relative to what was known from experiments. In this work, we present transferable coarse-grained (CG) models of carbohydrates that can accurately predict their structural, chemical, and physical properties. To comprehensively assess and enhance the model's robustness by estimating uncertainty in both parameter and property predictions, we developed and employed a Bayesian Uncertainty Quantification (BUQ) framework. First, we used the Sobol sequence to perturb the final set of CG FF parameters and performed MD simulations to generate training data. This data was used to train a Gaussian Process Regression (GPR) model, which was used as a surrogate model during BUQ sampling. The BUQ sampling was performed using the Ensemble Slice Sampling (ESS) algorithm. The data obtained from the posterior distribution of parameters and properties was then used to evaluate the robustness of the parameters. Currently, we are employing these models to design novel glycomaterials for biomedical applications.

248b Variable Reaction Coordinate Transition State Theory for Computing the Rate Constant of Barrierless Reactions on Catalytic Surfaces

Youbin Kim, C Franklin Goldsmith

Harmonic Transition State Theory (HTST) has been a cornerstone in predicting rate constants for chemical reactions with first-order saddle points. Despite its widespread application and efficiency, it encounters limitations in accurately determining reaction rates for reactions without the first-order saddle point (also known as barrierless reaction), such as homolysis in the gas phase. To address these constraints, Variational Transition State Theory (VTST) was developed. VTST significantly enhances our understanding of chemical reaction mechanisms and improves our ability to predict reaction rates of those reactions without the saddle point. Distinguished by its various forms, each offering unique benefits, VTST encompasses a range of methodologies. This research primarily examines the Variable Reaction Coordinate Transition State Theory (VRC-TST), a method that excels in the evaluation of multi-dimensional partition functions for the transition state. We can use VRC-TST to analyze the kinetics of barrierless reactions accurately. Here, we delve into the core principles of VRC-TST and introduce the open-source software, ROTDpy. This software represents a significant advancement, facilitating the precise calculation of rate constants for a wide range of reactions. Although the method was developed originally for radical-radical recombination in the gas phase, we have now extended this methodology to include barrierless adsorption desorption reactions. In this study, we have employed ROTDpy to compute the rate constants for some systems, including the barrierless desorption of carbon monoxide and nitrogen monoxide from various transition metal surfaces. The results from these computations are compared with the Potential of Mean Force approach of Doren and Tully, thereby providing a robust validation of our computational findings. [1,2]

[1] Doren D. J., Tully J. C., Dynamics of precursor-mediated chemisorption. *J. Chem. Phys.* ***94***, 8428 (1991).

[2] Doren D. J., Tully J. C., Precursor-mediated adsorption and desorption: A theoretical analysis. *Langmuir* ***4***, 256 (1988).

248c Efficient Sampling of Many-Body Coarse-Grained Potential of Mean Force for Polymer-Grafted Nanoparticles through Non-Equilibrium Forward-Reverse Path Integral.

Shih Kuang Lee, Melody Zhang, Rebecca Lindsey, Sharon C. Glotzer

Polymer-grafted nanoparticles (PGNs) can self-assemble superlattices with intricate crystal symmetries beyond simple crystal structures such as simple cubic or face-centered cubic crystals. Alongside the recent advancements in synthesizing these PGNs to understand their complex crystal assembly, computer simulations pave the way for comprehensively investigating assembly of these complex crystal systems. However, PGNs have a large number of associated high degrees of freedom, making simulation studies of the sluggish assembly kinetics computationally expensive. The Coarse-Graining (CG) method mitigates this issue by treating a nanoparticle and its surrounding polymer complex environment as a single "virtual-particle", greatly reducing the degrees of freedom by treating inter-particle interactions via a Coarse-Grained Potential of Mean Force (CGPMF). Yet, the non-negligible many-body effect emerging at the thick grafting layer regime greatly complicates the coarse graining process. In this study, we combine techniques introduced in Kostzin et al. [1] and Zhou et al. [2] to enable computationally efficient generation of robust CGPMFs for PNGs. Our approach entails non-equilibrium forward-reverse path integral sampling with carefully chosen reaction coordinates, through the steered molecular dynamics (SMD). We validate our technique by comparing resulting two-body PMFs against those obtained from umbrella sampling method. Furthermore, we demonstrate the calculation of two and three-body CGPMFs, which, when combined, approximate the full many body CGPMFs. Additionally, our technique can be applied to consider up to four-body CGPMFs. Ultimately, our technique offers an efficient and reliable route for coarse-graining the PGNs, thereby extending the achievable simulation timescales.

[1] Kosztin, Ioan, Bogdan Barz, and Lorant Janosi. 'Calculating potentials of mean force and diffusion coefficients from nonequilibrium processes without Jarzynski's equality.' *The Journal of chemical physics* 124, no. 6 (2006).

[2] Zhou, Yilong, Sigbjørn Løland Bore, Andrea R. Tao, Francesco Paesani, and Gaurav Arya. 'Many-body potential for simulating the self-assembly of polymer-grafted nanoparticles in a polymer matrix.' *npj Computational Materials* 9, no. 1 (2023): 224.

248d Approximating Anisotropic Pairwise Potential Energy Surfaces Using Mixed Basis Surrogate Models with Sparse Sampling

Mohammadreza Fakhraei, Michaela Bush, Michael Howard, Christopher Kieslich

Accurately approximating anisotropic interactions poses a significant mathematical and computational challenge because these interactions depend on up to six configurational degrees of freedom. Surrogate models trained on limited data sampled on a predefined sparse grid in configuration space are a promising strategy to address this challenge. Here, we design surrogate models for anisotropic pairwise interactions using mixed basis interpolants on Smolyak sparse grids. We first apply a physics-informed coordinate transformation that significantly improves approximation accuracy. This transformation naturally identifies aperiodic and periodic configuration coordinates, which we represent using a basis of Chebyshev polynomials and trigonometric functions, respectively. We test our approach for different physical problems, including the interactions between dipoles and between shape-anisotropic nanoparticles. We also compare our strategy to state-of-the-art machine-learning alternatives and compare accuracies as a function of amount of sampled data. We anticipate that our approach will be generally useful for approximating anisotropic potentials of mean force and torque for mesoscale simulations.

248e Evaluating Machine-Learned Interatomic Models through Graph-Based Descriptor Analysis

Benjamin Laubach, Rebecca Lindsey

Atomic-scale simulations are vital for advancing chemical and materials research, providing intricate, experimentally inaccessible mechanistic and kinetic insights. However, their effectiveness hinges on underlying interatomic models (IAMs) used to describe the system's potential energy surface. Quantum-based methods offer high accuracy but are computationally demanding for systems larger than a few hundred atoms whereas classical "force field" methods are orders of magnitude more computationally efficient but have limited accuracy. Machine-learned interatomic models (ML-IAM) provide a powerful means of bridging this accuracy and efficiency gap by learning suitable IAM from sufficient training data. However, the development of IAMs face hurdles including the need for large informative training sets, improved automation, and uncertainty quantification. A lack of suitable metrics for quantitative description and comparison of simulation configurations are critical for addressing these challenges. In this presentation, we describe efforts to develop innovative physics informed graph-based "fingerprinting" methods to fill this capability gap. Notably, our approach provides insight into how ML-IAMs perceive system structure and lends new physical insights into particularly complex chemical and material systems.

248f A Simple Semi-Explicit Statmech Model of Water That Captures Dynamic Solvation Interface

Lakshmanji Verma, Ken A. Dill

Molecular simulations in aqueous solution rely on either explicit water, which is computationally expensive, or implicit water, which can miss much of the interfacial physics. Here we develop a statistical mechanical model of liquid water and its interface with solutes. It has the advantages of being efficient -- because it is a computation, not a simulation -- and of capturing water's tetrahedral hydrogen bonding and spherical Lennard Jones interactions. It entails two steps. First, in *Cage Water,* we develop a reference state of pure water over the full liquid range vs (T, p) [Urbic, Dill, JACS, 2018]. Second, in *Crust Water* [Yadav et al, JPCB, 2022], we introduce a spherical solute molecule and populate the solute's surface (i.e. its *crust*) with these semi-explicit water molecules that we can compute efficiently. Now we further develop this model, more physically and accurately, by capturing the translational symmetry of dynamic water molecules in the first solvation shell and by introducing ions to predict solvation of charged molecules more faithfully. We think that further developments will bring this approach into computational modeling of large scale physical and biophysical simulations.

248g Finite-Size Effects and Structure-Selectivity Relationships in Ion Transport through Nanopores

Omar Khalifa, Brian Shoemaker, Amir Haji-Akbari

Understanding ion transport through nanopores is crucial for various separation processes, such as metal recovery, water desalination, and biological applications. The ultimate objective in membrane-based separations is to optimize the selectivity of targeted molecules while maintaining high permeability. Although achieving both goals simultaneously is challenging, efforts are made to minimize compromise. One approach involves testing new classes of materials with high permeability and selectivity, aiming to identify potential candidates for membrane design. Another approach is to optimize membrane design by exploring structure-selectivity relationships. Experimentally investigating such relationships at the scale of single-digit nanopores (SDNs) remains challenging due to limitations in achieving high spatiotemporal resolution. However, advancements in parallel programming algorithms, molecular simulations, and advanced sampling techniques have enabled access to very short time and length scales that were previously inaccessible through experiments. In this work, we investigate pressure-driven ion and water transport through graphene-based membranes (see figure) using molecular dynamics (MD) and forward flux sampling (FFS).

Modeling ion transport through membranes can introduce significant finite-size effects, particularly when the simulation box size is *inadequate*. Addressing these effects is crucial to ensure the validity and accuracy of simulations. In this study, we delve into the nature and extent of these finite-size effects arising from extraneous electrostatics imposed by the periodicity of the simulation box. Notably, these effects become pronounced when a charged particle (ion) traverses the membrane, leaving behind a charged feed. The periodic images of this feed exert artificial forces on the traversing ion. Furthermore, discrepancies in the dielectric constants across different system regions (feed, membrane, and filtrate) induce additional effects. To mitigate these challenges, we developed a physics-based model, specifically the ideal conductor model followed by a generalized model (Ideal conductor dielectric model - ICDM), which accounts for dielectric effects. This model effectively corrects for finite-size effects, enabling the derivation of accurate free energy profiles. Consequently, we can predict passage times and selectivities with improved precision. Additionally, we explore the extent of finite-size effects concerning pressure-driven hindered ion transport through nanopores in various systems characterized by structural factors such as the number of graphene layers and strength of pore charge. Our objective is to determine an optimal simulation box size that minimizes finite-size effects while maintaining computational efficiency. Implementing such a model enhances the fidelity of membrane-based simulations, aligning with the methodology employed in this study.

Through our investigation of multi-pore graphene membranes, we have uncovered significant pore-pore correlations that arise with proximity, exerting considerable influence on ion and solvent passage. Notably, these correlations adversely impact the passage of counter-ions and the polar solvent (water), while facilitating the passage of co-ions. As a result, both water permeability and salt rejection are diminished in close proximity with time scales that deviate from predictions made by models like the access resistance model. To address these findings, we propose a new, straightforward phenomenological model. This model considers pre-pore ion concentration and the forces exerted on traversing ions through dipole-decorated pores, enabling us to predict the relative impact of pore-pore correlations on ion selectivity in multi-pore membranes. Leveraging this model, we systematically screen multiple pore arrangements to identify those with minimal pore-pore correlations. Furthermore, our investigation delves into the effect of pore permanent dipole strength in terms of partial charges on ion selectivity and water permeability. Preliminary results suggest that counter-ions traverse faster through pores with stronger dipoles, while co-ions exhibit slower traversal due to reduced electrostatic pull from stronger charged dipoles. Overall, this work contributes to the design of more efficient membranes and enhances our understanding of nanoscale separation mechanisms, including those relevant to biological processes.

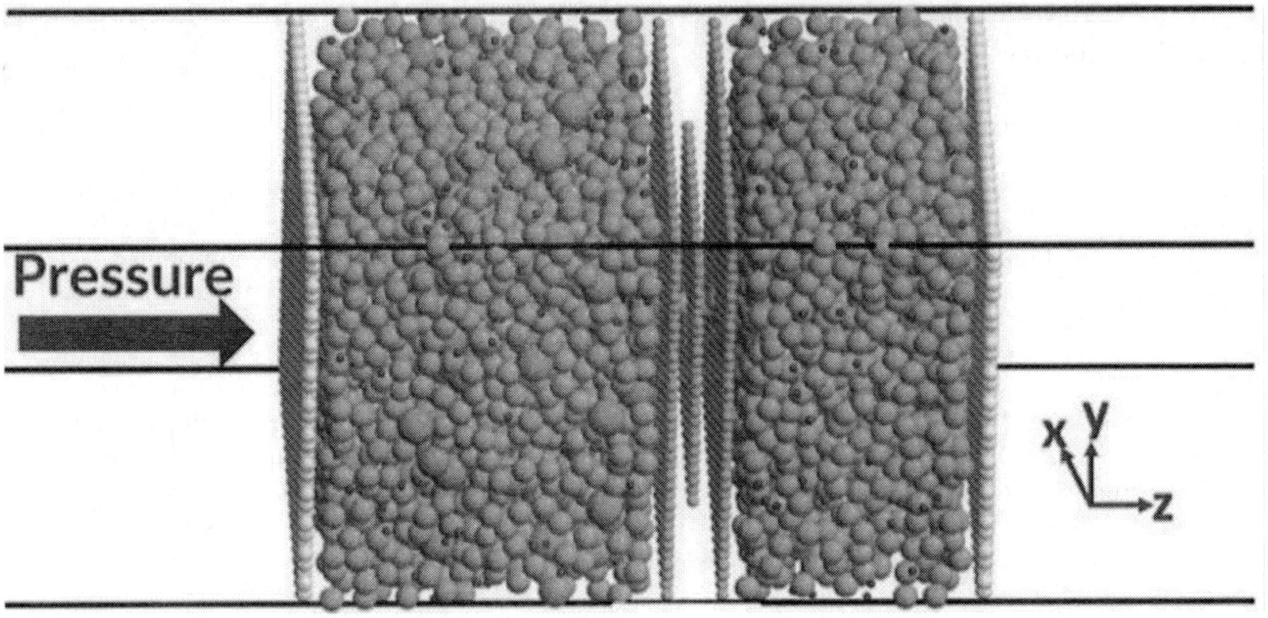
Pressure
x
y
z

248h Theory and Monte Carlo Simulation of Ideal Gas with Shell Particles in Canonical, Isothermal-Isobaric, Grand Canonical and Gibbs Ensembles

Harold Hatch, Vincent K. Shen, David Corti

Theories of small systems play an important role in the fundamental understanding of statistical mechanics, finite size effects and the validation of molecular simulations because no computer can simulate fluids in the thermodynamic limit. Previously, a shell particle in the isothermal-isobaric ensemble resolved ambiguity in the partition function, removed translationally redundant degrees of freedom and demonstrated quantitative differences from traditional simulations. In this work, shell particles are included in the canonical, grand canonical and Gibbs ensembles using ideal gas theory and Metropolis Monte Carlo simulation. We investigated the system size dependence of various ways of conducting ensemble averages. Furthermore, Gibbs ensemble simulations with shell particles removed a known instability that occurs without shell particles when a system reaches zero particles and infinitesimal volume. These expressions serve as a convenient test for Monte Carlo simulations.

248i Grand Canonical Approach to Modeling Dynamic Catalysts: From Thermal to Electro-Catalysis, from Clusters to Surfaces

Zisheng Zhang, Philippe Sautet, Anastassia Alexandrova

Dynamic structural rearrangement has been observed in a wide range of heterogeneous catalysts and functional materials when they are in operation. Such fluxional behaviours underlie the reactivity, activation, or deactivation in various catalytic systems. However, experimentally resolving their atomic structures has also been challenging due to the transient and minority nature of the metastable motifs and surface phases. The role of theory in investigating those dynamic systems hence remains unitary.

This talk will focus on the development and application of a grand canonical (GC) approach to model catalysts that undergo significant off-stoichiometric restructurings in reaction conditions. Grand canonical genetic algorithm (GCGA), an efficient global optimization algorithm, is implemented and used to explore the vast chemical space of cluster isomerization, surface atoms rearrangement, mixed coverage and configuration of adsorbates, and locate the global and relevant local minima. The found minima constitute a GC ensemble of catalyst states that are diverse in structure, stoichiometry, and reactivity. By thermodynamics and GC-DFT calculations, the dependence on reaction conditions (temperature, partial pressures, pH, electrode potential, solute concentrations, etc.) can be included into the free energetics of the states, to probe how the distribution of states responds to varying conditions.

This approach has been applied to investigate multiple systems ranging from thermal to electro-catalysis, and from supported clusters to extended surfaces. This talk will cover a few representative systems, including boron nitride in thermal oxidative dehydrogenation conditions, supported sub-nanometer metal clusters in electrocatalysis, and copper electrodes in electroreduction conditions. The collection of works will illustrate how the GC ensemble approach not only helps interpret complex experimental observations, but also provides rich atomic insights into the structure and reactivity of catalytic species, and lays the foundation to build a new paradigm for reaction kinetics, catalyst optimization, non-equilibrium behaviors, and more.

248j Overturning Intrinsic Temporal Limits in Atomic Scale Simulation of Reactive Systems

Dylan Gilley, Vignesh Sathyaseelan, Brett Savoie

Myriad research topics in chemical engineering and materials science belong to the category of "reactive systems." Fundamental understanding of these systems is achieved most definitively and most efficiently with atomic level modeling. State-of-the-art methods based in Molecular Dynamics (MD) such as ab initio MD (AIMD), reactive MD, and hybrid MD/Kinetic Monte Carlo (HMD/KMC) provide such spatial resolution. However, simulation past nanosecond timescales is not possible due to the significant computational demand. Here, a new HMD/KMC algorithm is introduced that is capable of atomistic simulation of reactive systems to arbitrarily long timescales. This is achieved by placing the burden of timekeeping on the KMC portion of the methodology, as opposed to the current strategy of using MD for the official timeclock. Despite being heuristic, this algorithm has been used to accurately reproduce dynamic and equilibrium behavior of multiple test systems. Homogeneous gas- and liquid-phase systems provide facile calculation of reference data through the application of kinetic theory. Accordingly, the HMD/KMC method shows quantitative agreement of molecular concentrations across timescales previously achievable only through continuum modeling (e.g. tens of seconds). This algorithm can be used to explain phenomena of reactive systems for which experimentation and continuum modeling cannot provide definitive answers. For example, device failure in lithium-ion batteries is acutely sensitive to degradation of the Solid Electrolyte Interphase (SEI). Due to its highly heterogeneous structure and slow degradation reactions, neither continuum models nor current atomistic simulations can readily describe these processes. Simultaneously, these phenomena occur at the nanoscale over significant periods of time, making experimental measurement difficult and expensive, at best. With this algorithm, an accurate description at the appropriate spatial resolution will provide crucial insights into how and why the SEI degrades. This talk will discuss the structure and capabilities of this HMD/KMC algorithm. Through presentation of the validation systems, it will be shown that quantitatively accurate atomistic simulations of reactive systems are no longer restricted to short time scales.

248k Gas Phase Reaction Exploration with Reactive Active Learning

Siddarth Achar, Priyanka Bholanath Shukla, Chinmay Mhatre, Caitlyn Vinger, Leonardo Bernasconi, Karl Johnson

Ab initio methods have limitations in studying chemical reactions due to their computational costs. Empirical forcefields are computationally inexpensive but lack bonding terms to model chemical reactions. Bond-order potentials like ReaxFF enable the sampling of bond-breaking and bond-formation within a system. However, these forcefields lack accuracy and require reparameterization for new systems. Machine learning interatomic potentials (MLIPs) can be used to explore numerous reactions at costs comparable to empirical forcefield methods and with near-ab initio accuracy. However, MLIPs are as good at sampling chemical reactions as the data that they are trained on. Data generation schemes that involve performing long MLIP molecular dynamics (MD) simulations are inefficient to enumerate reactive events. We tackle these challenges by formulating an active learning scheme that uses MLIPs and transition-state finding techniques to generate highly informative reactive datasets. This scheme does not require information of reaction pathways or knowledge of final products, thus minimizing human bias. Our MLIPs are trained using the DeePMD formalism with this reactive active learning scheme. We have accurately determined reaction pathways and transition state barriers for gas phase ammonia synthesis and the reaction of methylene imine with water molecules. We found our method to be data efficient as compared to conventional MLIP active learning schemes. Our scheme can determine new minimum energy pathways for various reactions, making it a robust tool for efficiently screening reactions in small molecule design and discovery.

248l Data-Driven Design and Identification of Active Site Motifs in Amorphous Porous Polymer Catalysts Using Molecular Simulation and Reinforcement Learning

Ryan Zhou, Brandon C. Bukowski

Amorphous porous organic polymers (POPs) offer significant design versatility for specific catalytic applications due to their flexible and shape-selective nanopore environments compared to zeolites and Metal-Organic Frameworks (MOFs). In particular, POPs functionalized with acid sites can catalyze esterification reactions[1] and selective organic conversions, including the formation of cyclic ester carbonates from CO_2[2]. To effectively evaluate the catalytic performance of a POP computationally, its structure must be simulated, active sites motifs identified, and after which Density Functional Theory (DFT) can be used to calculate the adsorption or activation energies. The intricate and amorphous nature of POPs complicates this conventional computational catalysis workflow, resulting in challenges for feature engineering and high computational costs for high-throughput screening. We introduce a strategy comprising two cooperative methodologies to address these limitations. The first part of our strategy is applying a reinforcement learning (RL) algorithm to streamline the discovery and optimization of potential catalysts within POP networks. This approach distinguishes itself from conventional machine learning approaches that typically require extensive labeled databases and may not effectively handle the dynamism and complexity in catalyst design. Our approach leverages RL to efficiently navigate the design of POPs to generate polymers with desired catalytic properties. We generated structures by simulating the polymerization process using Polymatic[3], in which monomers were packed and bonded using molecular dynamics to create an amorphous polymer structure. We then used RL to modify polymerization conditions and assess the obtained pore environments. The second part of our strategy uses image recognition and clustering tools to identify amorphous active site motifs based on their local environments generate surrogate models that are tractable with Density Functional Theory (DFT). Our ML model integrates these two components for high-throughput structure generation and screening. By prioritizing kinetically relevant regions within the polymer networks, our AI model improves the efficiency of identifying potential active sites in POPs.

Our RL environment training results demonstrate the capability to improve the adsorption properties of polymers and to target specific catalytic properties as training objectives. In a simulation environment using styrene, divinylbenzene, or porous aromatic frameworks (PAFs) as example initial monomers at different mixing ratios[4], our model aligns its learning trajectory towards improving desired textural properties, such as average pore size, pore-limiting diameter, and the completion rate of polymerization, by setting these textural features as goals. For identification of active site motifs, our AI model incorporates the screening workflow that combines Density Functional Theory (DFT) and Monte Carlo simulations, streamlined for efficiently computation of applications such as CO2 cyclization and ammonia decomposition. Tools such as PoreBlazer are implemented in our model for analyzing structural properties, offering a comprehensive assessment of catalytic capabilities. This screening methodology enables our model to effectively improve and update the training policies. The results show the improvement of porous structure and catalytic capabilities of polymers simulated in an RL environment by the ML workflow, highlighting the efficacy of this approach in enhancing polymer design for catalysis and adsorptive separations.

629a Modeling Affinity Maturation to Recapitulate Germinal Center Dynamics and Resulting Antibody Sequences Against Sars-CoV2 and HIV

Jonathan Faris, Kayla G. Sprenger

Highly mutable infectious disease pathogens such as HIV and influenza evolve too rapidly for the human immune system to contain them, allowing them to circumvent traditional vaccination approaches, and cause over one million deaths annually. To combat these pathogens and others, we must develop tools to aid in the design of robust, potent, and universal immunogens. Agent-based models can be used to simulate the complex interactions and dynamics that occur in the lymph nodes between immune cells and pathogen-like proteins (antigens) during affinity maturation—the process by which antibodies evolve. As such, compared to existing experimental approaches, agent-based models offer a safe, low-cost, and rapid route to study the immune response to vaccines spanning a wide range of design variables. Previously, models of affinity maturation have been limited in their ability to leverage the sequence and structure of the antibody/antigen complex, replicate the immunological dynamics found *in vivo*, or both. Herein, we build upon our recent efforts incorporating antibody nucleotide sequences, antigen amino acid sequences, and 3D crystal structures. Through the development of both more physiologically relevant B cell dynamics and an explicit antigen capture model, we accurately recapitulate a variety of relevant metrics seen *in vivo*. Importantly, the model obtains these results without sacrificing the sequence or atomistic-level details of the evolving B cells. Preserving these important biological factors allows for direct examination of the resulting antibody sequences for clinically relevant clonotypes observed in the general population following exposure to antigens of interest via infection or vaccination. By accurately capturing both the dynamics of affinity maturation and the details of the resulting immune response, we believe this model will enable the rational design and preliminary screening of immunogen candidates for a variety of pathogens which have thus far proven too difficult or too expensive to test by traditional means.

629b Kinetics without Dynamics: Reactive Coarse-Graining with a Real Time Coordinate

Vignesh Sathyaseelan, Dylan Gilley, Brett Savoie

Lithium-ion batteries (LIB) are found everywhere from portable electronics to electric cars. The primary components of lithium-ion batteries are the graphite anode, the separator, an organic electrolyte, such as ethylene carbonate (EC) and propylene carbonate (PC), and the transition metal-oxide cathode. Another crucial component of the LIB that has a major impact on its stability and life span is a passivation layer that forms on the anode called the Solid-Electrolyte Interphase (SEI). The SEI imparts stability to the battery and its breakdown is a major cause of catastrophic degradation of the battery. However, there is limited understanding of the molecular-scale events that lead to the catastrophic failure of batteries.

Molecular Dynamics (MD) simulations are widely used to study systems at the molecular scale; however, MD simulations are unsuitable to study the molecular processes leading to battery failure as this is a highly reactive system and classical MD suffers from the inability to account for reactions that take place within the simulated system. Methods that can model reactive systems such as Ab-Initio Molecular Dynamics (AIMD) are prohibitively expensive to reach timescales that are relevant to battery degradation. Therefore, there is a need for computationally inexpensive methods that can be used for modeling the dynamics of reactive systems. One approach to solve this methodological gap is hybrid Monte Carlo/ Molecular Dynamics Methods (MC/MD). This method combines MD and MC, integrating MD simulations for transport processes and MC simulations for chemical reactions, our approach offers a comprehensive framework for modeling reactive systems in a computationally tractable manner.

Here, we report the development of the hybrid Kinetic Monte Carlo/ Molecular Dynamics (KMC/MD) method along with the validation of the method by studying the combustion of alkanes, such as methane. Initially targeting Solid-Electrolyte Interphase (SEI) formation kinetics presents challenges due to diffusion limitation making it a confounding factor for interpreting SEI formation kinetics. The complexity and unique challenges posed by methane combustion render it an ideal non-trivial benchmark for evaluating the reactive MD algorithm's efficacy. By focusing on methane combustion, we aim to validate the KMC/MD algorithm's ability to accurately capture the dynamics of reactive systems characterized by rapid chemical transformations and the generation of diverse molecular species. This preliminary validation step will provide crucial insights into the algorithm's performance and pave the way for its application to more complex systems, including SEI formation in Li-ion batteries.

For the benchmark problem, the reaction network was synthesized using an automated reaction network generation framework to find the possible products for a set of reactants and the transition state (TS) connecting the reactants and products along with the activation barriers using Density Functional Theory (DFT). Once a comprehensive list of reactions and the species involved in these reactions have been generated, force field (FF) parameters are required to capture the transport processes in the MD step. The commonly used FFs use experimental data during the parameterization process. This reliance on experimental data results in limited molecular coverage and impedes the study of reactive systems. To address this problem, a general-use FF parameterized purely on DFT calculations was also developed in this work. The reactive MD simulation itself consists of MD runs of constant time intervals followed by KMC steps to execute reactions. This alternating cycle of MC and MD continues until the maximum simulation time is reached. The hybrid KMC/MD algorithm incqorporates acceleration schemes such as selective downscaling of fast reactions that are in a pseudo-steady state (PSS) to enable the simulation of long timescales.

The resultant simulation outcomes are then compared against ground truth kinetics obtained from microkinetic modeling. Our analysis demonstrates a notable concordance between the kinetics profiles derived from the hybrid MD/MC simulations and those obtained from microkinetic modeling (MKM). This agreement persists across varying temperatures and initial concentration ratios, underscoring the algorithm's efficacy in capturing the dynamics of methane combustion systems. Furthermore, this work sets the initial groundwork for exploring the degradation of more complex reaction chemistries, particularly of the SEI degradation in Li-ion batteries.

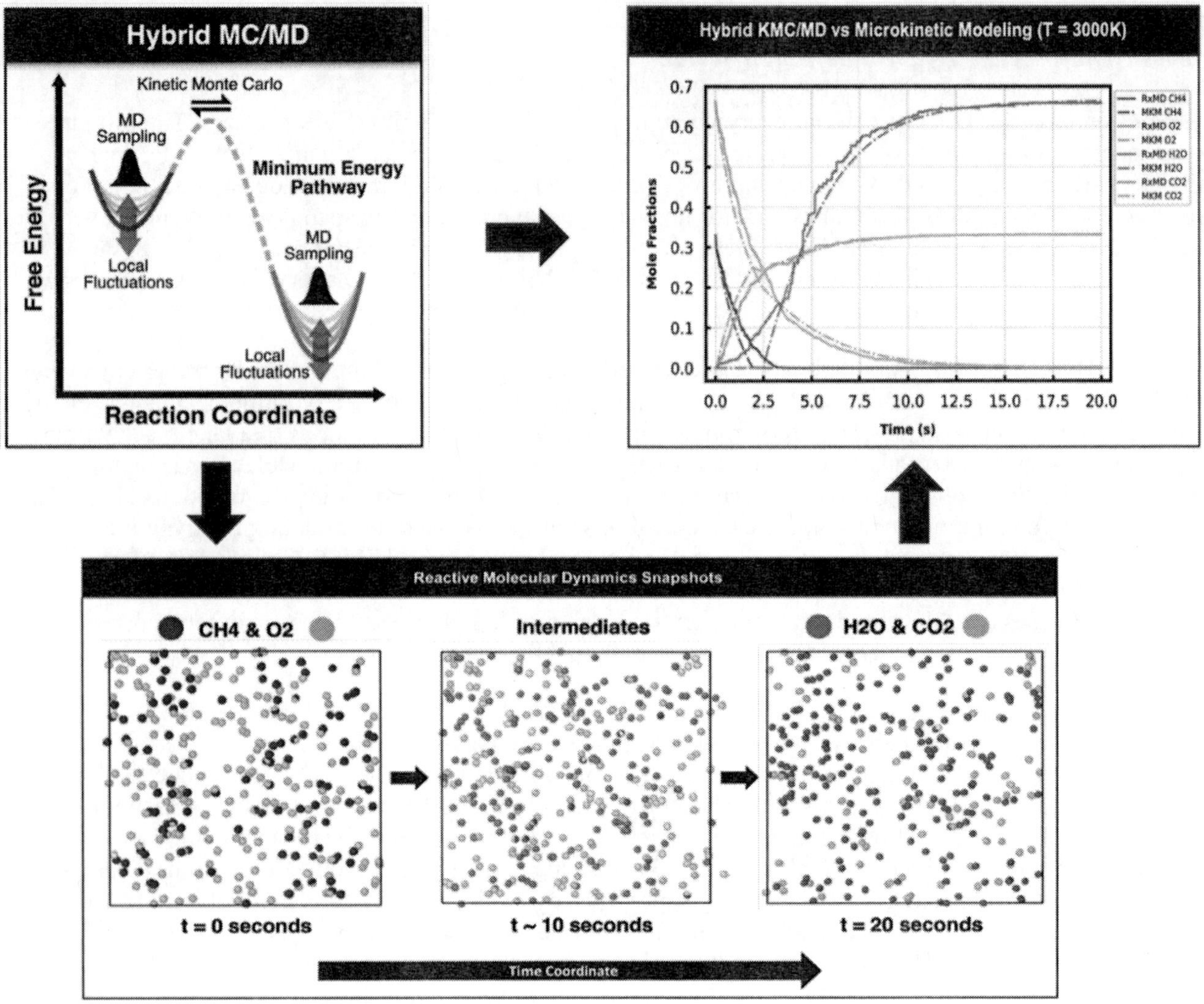

629c On Modeling the (De)Polymerization of Hyperbranched Polyesters

Matthew W. Coile, Alexander Shaw, Sai Phani Kumar Vangala, Changxia Shi, Eugene Y.-X Chen, Linda J. Broadbelt

Hundreds of millions of tonnes of plastic are produced each year, only a fraction of which is currently recycled. This enormous quantity of polymer waste represents both a significant environmental challenge as well as an opportunity to recover valuable chemicals or materials. Many of the polymers produced today have been optimized for properties relevant to their primary use, without regard to their ability to be depolymerized or recycled. Polyesters synthesized from lactones have recently shown promise as materials that can undergo depolymerization at their end-of-life. A particularly exciting hyperbranched polyester synthesized from a lactone monomer was shown to depolymerize with >90% conversion to monomer with undetectable side products. However, it remains unknown how the reaction kinetics depend on the structure of the polymer chains and the functional groups present on each chain unit. We developed a mechanistic kinetic model for the reversible growth of this hyperbranched polyester system to gain insight into the growth mechanism, polymer structure, and its depolymerization potential. Kinetic Monte Carlo (kMC) was used to track chain-specific detail, including the full structure of each polymer chain. Kinetic parameter estimates were derived from density functional theory (DFT) calculations. Transport limitations were accounted for by using an effective rate constant that was a function of both the intrinsic kinetics rate constant and a chain-length dependent transport rate constant, and it was found that this further improved the model's ability to capture the trends in the experimental data. The model was then used to probe characteristics of the system that are not readily explored by experiment itself, such as catalyst coverage and how the dominant growth mechanism shifts over the course of the reaction. This model lays the foundation for future design and study of hyperbranched polyesters derived from lactones, leading to new readily depolymerizable polymers.

629d *Mechanistic Insights into the Effect of Surface Charge Oscillation on Ammonia Synthesis from First Principles Kinetic Simulations*

Rajat Daga, Ulrick Gaillard, Matthew Neurock

Ammonia is one of top chemicals produced with applications in agricultural and energy storage industry as a hydrogen carrier alternative. While catalyst development has significantly improved ammonia yield over the past few decades, further advances have been limited as the metals used is controlled by the Sabatier's principle. Dynamic catalysis is a promising approach that may help to overcome this limitation. Recent efforts by Frisbie and Dauenhauer have reported on the development of "catalytic condensers" that enables facile change to the charge induced into supported catalysts[1,2,3]. Herein, we evaluate the efficacy of dynamic catalysis using surface charge oscillation on Ru (0001) surface to enhance ammonia yield. We employ a combined Density Functional Theory (DFT) calculation together with Kinetic Monte Carlo (kMC) simulations to provide mechanistic insights into the role of surface charge oscillations in carrying out molecular transformations. DFT calculations were carried out as a function of charge as well as surface coverage to determine the kinetic parameters of the elementary steps and the binding energies of the adsorbates on Ru (0001) surface. These parameters were then subsequently used to parametrize lateral interaction models and provide elementary kinetic database used in kMC simulations to simulate turnover frequencies (TOF) as a function of reaction conditions and charge. kMC simulations on the static Ru (0001) surface revealed that the TOF decreased with increasing negative surface charge due to the weakening of Ru-N bond, which results in increased N_2 activation barrier. The TOF increased from 0.43 $NH_3(\text{Ru atom})^{-1}s^{-1}$ on neutral Ru surface to 0.86 $NH_3(\text{Ru atom})^{-1}s^{-1}$ at +0.04e/Ru atom surface charge due to stronger Ru-N bond strength with increasing positive surface charge. However, upon increasing the surface charge beyond +0.04e/Ru atom, the Ru-N bond strengthens leading to difficult desorption of NH_3. Therefore, +0.04e/Ru atom is the Sabatier limit with respect to the surface charges on the static Ru (0001) surface. kMC simulations on a dynamic Ru surface with square wave oscillation function between -0.06e and +0.17e per Ru atom yield a TOF of 2.97 $NH_3(\text{Ru atom})^{-1}s^{-1}$, exceeding the Sabatier limit by approximately three times. Overall, the combination of DFT and kMC simulations helps to elucidate the effect of surface charge oscillation on ammonia synthesis reaction network and indicates that the Sabatier limit can be surpassed using dynamic catalysis and provides a direction for future experimental efforts.

References:

[1] *Wang, Y., Udyavara, S., Neurock, M. and Frisbie, C.D., 2019. Field effect modulation of electrocatalytic hydrogen evolution at back-gated two-dimensional MoS2 electrodes. Nano letters, 19(9), pp.6118-6123.*

[2] *Onn, T.M., Gathmann, S.R., Guo, S., Solanki, S.P.S., Walton, A., Page, B.J., Rojas, G., Neurock, M., Grabow, L.C., Mkhoyan, K.A. and Abdelrahman, O.A., 2022. Platinum graphene catalytic condenser for millisecond programmable metal surfaces. Journal of the American Chemical Society, 144(48), pp.22113-22127.*

[3] *Onn, T.M., Gathmann, S.R., Wang, Y., Patel, R., Guo, S., Chen, H., Soeherman, J.K., Christopher, P., Rojas, G., Mkhoyan, K.A. and Neurock, M., 2022. Alumina graphene catalytic condenser for programmable solid acids. JACS Au, 2(5), pp.1123-1133.*

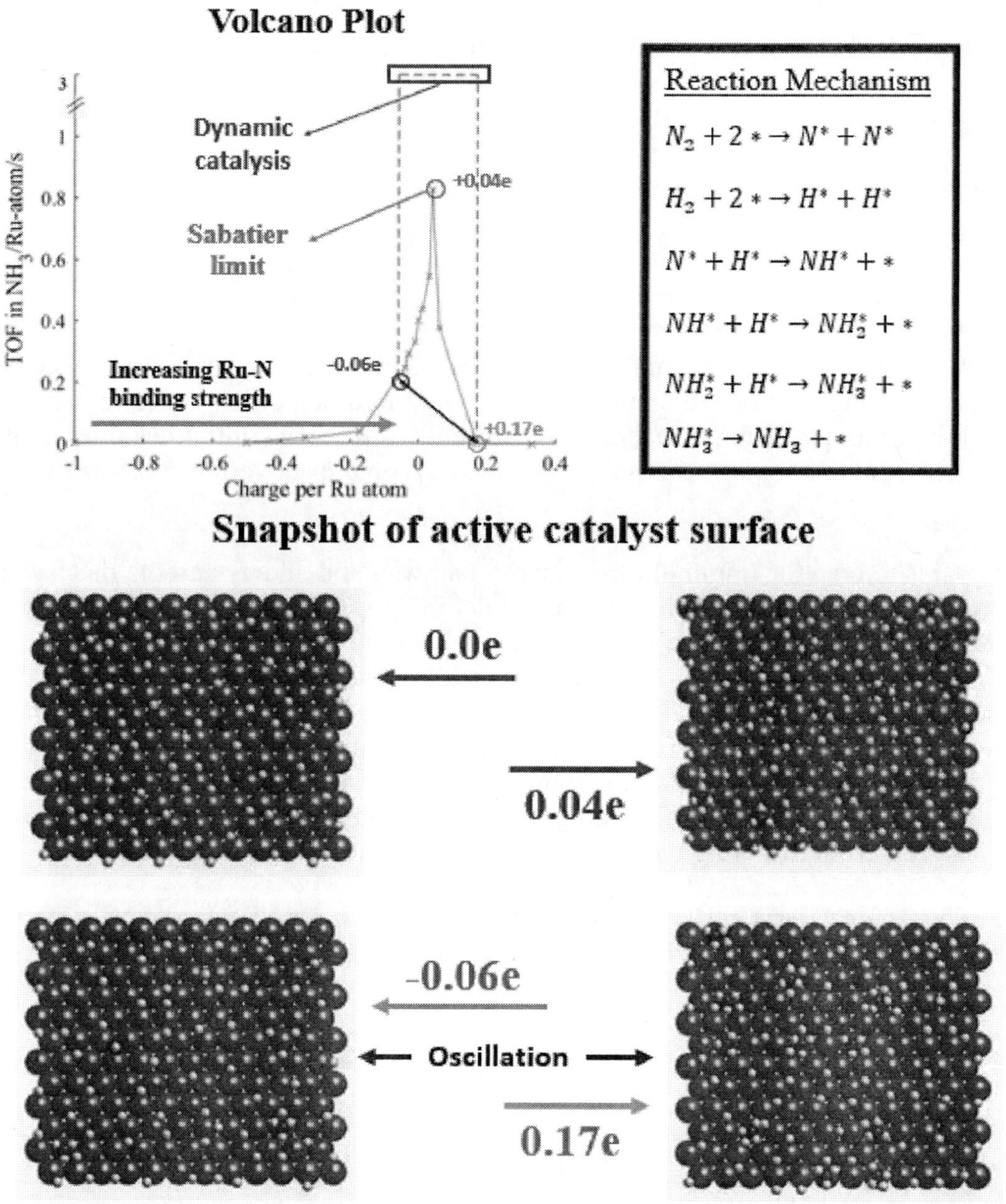

Volcano Plot
Dynamic catalysis
Sabatier limit
+0.04e
-0.06e
+0.17e
Increasing Ru-N binding strength
TOF in NH3/Ru-atom/s
Charge per Ru atom
Reaction Mechanism
$N_2 + 2* \rightarrow N^* + N^*$
$H_2 + 2* \rightarrow H^* + H^*$
$N^* + H^* \rightarrow NH^* + *$
$NH^* + H^* \rightarrow NH_2^* + *$
$NH_2^* + H^* \rightarrow NH_3^* + *$
$NH_3^* \rightarrow NH_3 + *$
Snapshot of active catalyst surface
0.0e
0.04e
-0.06e
Oscillation
0.17e

629e Resolution of the Minimum Free Energy Pathway of Integrin Activation Using Generative Modeling and Finite Temperature String Method

Siva Dasetty, Robert Coffman, Tamara C. Bidone, Andrew Ferguson

Understanding integrin activation is of large interest because of its central role in various cellular functions and in designing therapeutics targeting integrin.[1-5] However, accessing the full continuous conformations of integrin activation is challenging using experimental methods because of their transient nature and via all-atom molecular dynamics simulations because of the inordinate computational resources required as the system comprises more than a million atoms. To this end, we have previously developed a multi-scale simulation technique combining non-linear manifold learning and deep generative modeling methods to predict the full continuous activation conformations of integrin.[6] While this technique enabled access to the currently unknown continuous intermediate conformations of integrin activation, the thermodynamically favorable activation mechanism remains elusive.

In this work, our goal is to map the minimum free energy pathway and understand the mechanism of integrin activation by focusing on αIIbβ3 integrin, a platelet integrin whose malfunction can lead to bleeding disorders and cancer. For this, we employ finite temperature string method with conformations of the initial images generated by our generative modeling technique. We use the experimentally hypothesized integrin activation pathways to set the location of the initial string. In each string iteration, we perform all-atom molecular dynamics simulations with explicit water representation at biophysical conditions of each image in parallel to evolve the string. In our talk, we will present our generalizable multi-scale simulation technique along with the biophysical insights of integrin activation gained from resolving its minimum free energy activation pathway.

[1] Bidone TC, Polley A, Jin J, Driscoll T, Iwamoto DV, Calderwood DA, Schwartz MA, Voth GA. Coarse-grained simulation of full-length integrin activation. Biophysical journal. 2019 Mar 19;116(6):1000-10.

[2] Banno A, Ginsberg MH. Integrin activation. Biochemical Society Transactions. 2008 Apr 1;36(2):229-34.

[3] Zhang Y, Wang H. Integrin signalling and function in immune cells. Immunology. 2012 Apr;135(4):268-75.

[4] Goodman SL, Picard M. Integrins as therapeutic targets. Trends in pharmacological sciences. 2012 Jul 1;33(7):405-12.

[5] Tong D, Soley N, Kolasangiani R, Schwartz MA, Bidone TC. αIIbβ3 integrin intermediates: from molecular dynamics to adhesion assembly. Biophysical Journal. 2022 Dec 23.

[6] Dasetty S, Bidone TC, Ferguson AL. Data-driven prediction of αIIbβ3 integrin activation paths using manifold learning and deep generative modeling. Biophysical Journal. 2023 Dec 14.

629f Application of Graph Neural Network in Accurate Prediction of Force Chain Network in Dense Suspensions

Abhinendra Singh, Joao Maia, Armin Aminimajd

Concentrated or dense, packings of particulate systems in a liquid or air are commonly found in natural, human health, and industrial settings [1]. These systems range from blood to paint, concrete, etc. They are often out-of-equilibrium and display a wide array of rheological features, of which the liquid–solid phase transition under external deformation is arguably the most interesting. Over the last decade, research activity at the intersection of fluid mechanics, granular materials, driven disordered systems, and soft condensed matter physics has led to a consensus that the formation of mesoscale stress-activated frictional contact network (FCN) drives these mechanisms [2,3]. Particle simulations that led to this concept have been successful in quantitatively reproducing the non-Newtonian behavior of thickening suspensions [4]. Inferring when and where these FCNs appear is of vital importance in physics, geology, engineering, and industry due to their impact on material properties. While traditional simulation methods are expensive and time-consuming, recent deep learning techniques were found to be a powerful tool to simulate and predict properties of dry granular materials, colloidal, and complex physical systems. Among these, graph neural network (GNN) has proven to be robust and fast, using nodes and edges to represent particles and their relationships. Herein, we train the deep graph convolutional neural network (GCNN) model using datasets obtained from lubrication flow discrete element modeling (LF-DEM) to accurately predict FCN in suspensions at different volume fractions, step strains, stresses, disorder degrees, stiffness coefficient and friction coefficient. Our machine learning model goes beyond recent research on prediction of contact forces in granular materials; it not only predicts FCN with higher accuracy but also interpolates and extrapolates to conditions far from control parameters. The method used in this study can be utilized for prediction of rheological and characterization of complex particulate systems in the future. Once extensive validation is performed, these systems have capabilities to speed up current simulation and continuum modeling efforts, which can contribute to extraterrestrial building efforts.

References:

1. Morris, J. F. Shear Thickening of Concentrated Suspensions: Recent Developments and Relation to Other Phenomena. *Annu. Rev. Fluid Mech.* **52**, 121–144 (2020).
2. Singh, A. Hidden hierarchy in the rheology of dense suspensions. MRS Communications. **13** (6), 971-979 (2023).
3. M van der Naald, A Singh, TT Eid, K Tang, JJ de Pablo, HM Jaeger. Minimally rigid clusters in dense suspension flow, Nature Physics, 1-7, 2024.
4. Singh, A., Jackson, G. L., van der Naald, M., de Pablo, J. J. & Jaeger, H. M. Stress-activated Constraints in Dense Suspension Rheology. Physical Review Fluids **7** (5), 054302 (2022).

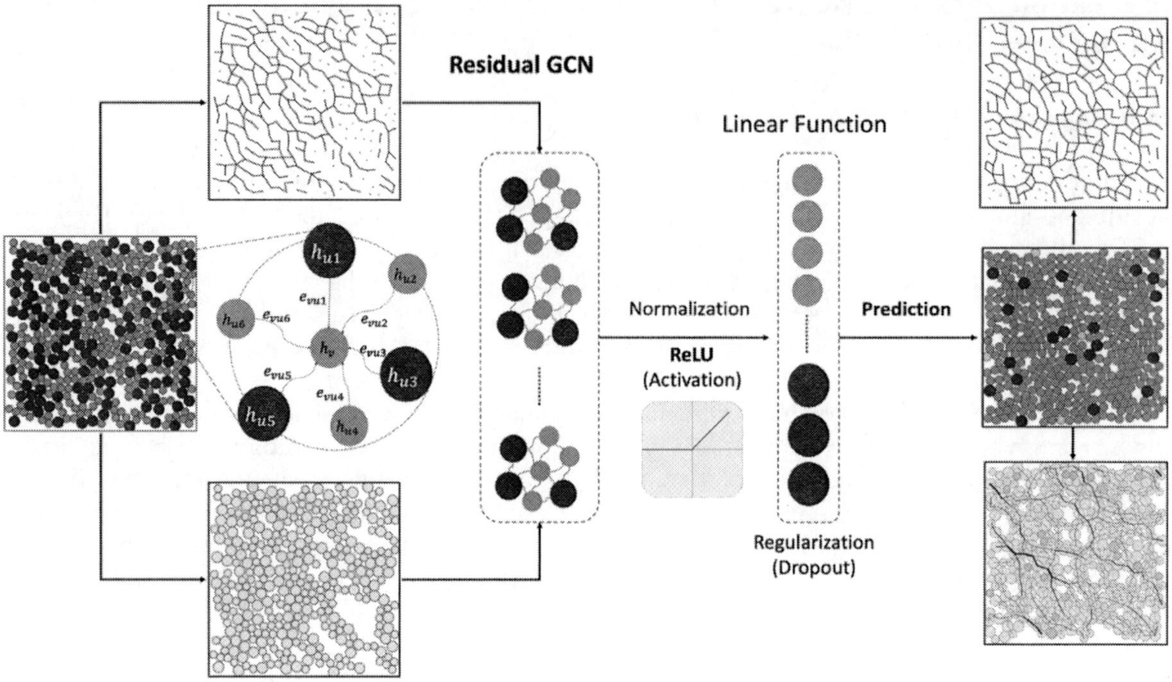

Residual GCN
Linear Function
Normalization
ReLU
(Activation)
Prediction
Regularization
(Dropout)

629g Holistic Framework for Coarse-Graining and Back-Mapping through Active Subspace Learning

Anna Wojnar, Logan Pike, Leah Bandy, Stephen Pankavich, Alexander J. Pak

Multiscale modeling and simulations can provide microscopic insights into emergent processes, such as those observed during soft matter or biomacromolecular assembly, that span multiple spatial and temporal scales difficult to study through experiments alone. Bottom-up coarse-graining is one such technique where information from high-resolution (e.g., atomistic) models is used to derive low-resolution (e.g., coarse-grained) models following statistical mechanical principles. Over the past decade, many algorithms to determine coarse-grained mappings and effective interactions have been proposed, and more recently, for the reverse process called back-mapping where high-resolution information is predicted from low-resolution data. However, these techniques have yet to realize their full potential for complex soft matter simulations, which we attribute to the fragmented (and sometimes *ad hoc*) nature of existing methods. In this work, we propose a new coarse-graining framework that leverages a supervised dimensional reduction technique called active subspace learning. We show that our approach is able to systematically determine coarse-grained mappings, effective interactions, and equations of motion while also facilitating back-mapping when combined with generative machine learning models. We demonstrate the holistic nature of our active subspace coarse-graining (ASCG) method through several examples of small proteins, where we assess both recapitulation of configurational probabilities and dynamical metrics. Finally, we discuss anticipated improvements to the ASCG method and its application to more complex biomolecular and soft matter systems.

629h Coulomb without Ewald: Accelerating Charged Molecular Dynamics with Field-Theoretic Simulations

Joshua Lequieu

In typical molecular simulations, the evaluation of long-ranged Coulombic interactions are solved using Ewald-based techniques. Yet despite the widespread use of Ewald-based methods, they remain computationally expensive and typically dominate the time required to evaluate the energies and forces in a simulation. In this work, we present an alternative approach for accelerating charged molecular dynamics using field-theoretic simulations. In contrast to particle-based simulations, the Coulomb operator in field-theoretic simulations can be evaluated locally, thereby permitting extremely efficient solutions of charged interactions without the need for costly Ewald-based techniques. Our approach leverages this advantageous feature to substantially accelerate molecular dynamics simulations involving charges. To demonstrate our approach, we show that charged particle and field-theoretic simulations can be constructed to be formally equivalent and yield identical results for both the pressure and chemical potential. Next, we compare the performance of these two methods and show that charged field-theoretic simulations can be three orders of magnitude faster than molecular dynamics simulations despite yielding equivalent results. We lastly illustrate our method for two brief examples: (1) the macrophase separation of oppositely charged polyelectrolytes and (2) the microphase separation of complex coacervate core micelles. Taken together, our work demonstrates that field-theoretic simulations can considerably accelerate calculations involving Coulombic interactions and can enable the study of new systems that are intractable with existing methods.

629i Transferable Protein Sidechain Backmappings Amenable to Reweighting

Jacob Monroe

While multiscale modeling of biomolecules has become a critical component in exploring their structure and self-assembly, backmapping from coarse-grained (CG) to fine-grained (FG), or atomistic, representations presents a challenge, despite recent advances through machine learning. A major hurdle, especially for those strategies utilizing machine learning, is that most backmappings can only approximately recover the atomistic ensemble of interest. We demonstrate the conditions for which backmapped configurations may be reweighted to exactly recover the desired atomistic ensemble. By then training separate decoding models for each sidechain type, we develop an algorithm based on conditional normalizing flows and geometric algebra attention that autoregressively proposes backmapped configurations. Critical for reweighting, our trained models include all hydrogen atoms in the backmapping and make probabilities associated with atomistic configurations directly accessible. Further, the models are sensitive to a sidechain's local environment and are transferable to any protein sequence. We anticipate that our models will be particularly useful for proposing Monte Carlo moves for simultaneously rearranging entire sidechains, or sets of sidechains, in addition to producing novel all-atom full protein configurations from CG models. We also demonstrate, however, that reweighting is extremely challenging despite state-of-the-art performance on recently developed metrics and generation of configurations with low energies in atomistic protein force fields. Through detailed analysis of configurational weights, we demonstrate that machine-learned backmappings must not only generate configurations with reasonable energies, but also correctly assign relative probabilities under the generative model. These are broadly important considerations in generative modeling of atomistic molecular configurations.

629j Flexible Membrane Protein Docking and Design with Biologically Realistic Implicit Membrane Models.

Rituparna Samanta

Membrane proteins participate in several life processes including transport, signaling and catalysis. They constitute over 30% of all proteins and are targets for over 60% of the FDA approved drug targets. Despite their importance, difficulties in overexpression and purification have limited the quality and quantity of experimental data, thus computational tools can help expedite the discovery. However, a critical limitation in capturing the complex environment results in either a too computationally expensive model or a severely simplified bilayer representation.

In this work, we will introduce an implicit approach that captures realistic features of the membrane such as the anisotropic density of water in the membrane, electrostatics due to the lipid layer and nanoscale dimensions of membranes with different lipid compositions. Due to the implicit nature of the model, our platform can tackle problems such as docking of larger proteins and rationally design a membrane protein given a backbone. Using two case studies, one on flexible docking of membrane proteins and another on designing beta barrel proteins, we will discuss improvements of this model and compare its performance with previous membrane models. Our method advances high-resolution membrane protein structure prediction and design toward tackling key biological questions and engineering challenges.

11a Efficient Implementation of Monte Carlo Algorithms on Graphical Processing Units for Simulation of Adsorption in Porous Materials

Zhao Li, Kaihang Shi, David Dubbeldam, Mark Dewing, Christopher J. Knight, Alvaro Vazquez Mayagoitia, Randall Snurr

In this study, we present enhancements in Monte Carlo (MC) simulation speed and functionality within a new molecular simulation code, gRASPA, that runs on graphical processing units (GPUs). The code achieves significant performance improvements over serial versions of the RASPA code through Nvidia MPS utilization and an option for scalable high-throughput screening. The gRASPA code supports a wide range of MC simulations, including grand canonical Monte Carlo, NVT-Gibbs, Widom test particle insertions, and continuous-fractional component Monte Carlo, with plans for ongoing feature addition and performance optimization. The code can also perform grand canonical transition matrix Monte Carlo (GC-TMMC) simulations for rapid free energy calculations. Our work also integrates machine learning (ML) potentials within gRASPA, and as an example, we present results for the adsorption of CO_2 in Mg-MOF-74, where the ML potential provides more accurate results than a classical force field. A broader audience can also benefit from the translated gRASPA code from CUDA to SYCL that can run on non-Nvidia GPUs. The gRASPA code is open source aiming to give access to the broader simulation community, who may also make future contributions to the ongoing development of this simulation framework.

11b Novel Tools for Studying Crystallization Pathways of Shaped Particles into Complex Crystals

Domagoj Fijan, Brandon Butler, Tommy Waltmann, Philipp Schönhöfer, Charlotte Zhao, Maria Ward Rashidi, Joshua Anderson, Sharon C. Glotzer

Nucleation and crystallization processes are critical in diverse materials, such as colloidal, soft, molecular, metallic, and ionic crystals. While there's considerable understanding of nucleation in simple crystals, research on complex crystals' nucleation is limited due to the lack of suitable computational tools. We address three main areas for improvement: an order parameter that describes local structures with continuous point group symmetry, a new neighbor list for shaped particles, and a tool for efficiently detecting and recording nucleation events.

Complex crystals exhibit multiple local environments, and to truly understand the nucleation process on a lowest level would require us to somehow track the progression of symmetry elements for each of these environments. However, traditional order parameters such as Steinhardt's (SOP) and its variants merge multiple symmetries into a single value, limiting their effectiveness in detailed analysis. Our novel Point Group Order Parameter (PGOP) overcomes this limitation by providing a per-particle order parameter based on symmetry operations. PGOP compares a particle's bond order diagram (BOD) with a perfect symmetrized BOD using the Pearson correlation coefficient. This quantifies how well the local environment aligns with specific symmetry operations, allowing distinct differentiation of local environments in complex crystals.

The accuracy of PGOP depends on identifying neighboring particles. Traditional methods like Voronoi tessellation and Solid-angle nearest neighbors (SANN) often fail in systems with shaped particles. To resolve this, we introduce Shadow Projection Overlap of Obstacles for Neighbor Exclusion (SPOONE). SPOONE identifies neighbors by casting rays and determining intersections (blocking), providing more accurate neighbor lists for shaped particles and enhancing PGOP's effectiveness. SPOONE can be utilized as a traditional or weighted neighbor list by considering the amount of blockage (shadow). This makes it very suitable in use for order parameter computation.

To facilitate the automatic detection of nucleation events and manage the data-intensive nature of these simulations, we developed Dupin, a Python tool for event detection in molecular simulation trajectories. Dupin promises to automate detection of events offering new potential applications of machine learning in the field. Dupin identifies changes using order parameters and employs change point detection methods to record events with high temporal resolution. Since dupin can work on-the-fly during the simulation it can also serve to minimizes data storage and improves simulation efficiency by selectively recording only significant events.

To showcase the utility of our newly developed tools, we present results from comparisons in traditional situations suitable for each tool. We contrast PGOP with SOP in both simple and complex crystals. SPOONE's effectiveness is demonstrated in systems of elongated shaped particles, with performance comparably effective for nearly spherical particles. Additionally, we explore the impact of different neighbor lists on PGOP and SOP by comparing the results using SPOONE versus Voronoi methods. Additionally, we highlight Dupin's ability to efficiently identify various events in soft matter simulations, including intricate nucleation processes and polymer collapses, for comprehensive post-simulation analysis. These tools collectively promise to advance the study of nucleation and crystallization, potentially opening new research opportunities in large-scale crystallization projects and enabling application of emerging fields such as machine learning in studying nucleation and crystallization pathways.

11c Relentless: Development of a Python Package for Transparent, Reproducible Simulation-Based Optimization of Materials

Chase Petix, Adithya N Sreenivasan, Zachary Sherman, Michael Howard

The ability to design a material with specific properties is one of the biggest goals of materials engineering, with potential to make breakthroughs in catalysis, separations, sensing, energy storage, and other important engineering areas. Computational inverse design is a promising general approach because it does not have the costs of laboratory experiments and can quickly explore design candidates. Many inverse design approaches incorporate molecular dynamics simulations as a component of a larger optimization workflow using proprietary code or "spaghetti" scripts that limit transparency and reproducibility. In this work, we present relentless, an open-source Python package that automates the optimization of objective functions computed with simulations. relentless carries out simulations using existing molecular dynamics software (HOOMD-blue, LAMMPS) to take full advantage of their optimized performance. A model, simulation protocol, and optimization strategy are specified through a high-level, object-oriented Python interface — making relentless easy to use, to extend, and to connect with other scientific Python packages.

11d Self-Assembly of Non-Spherical Colloids Using Coarse-Grained Simulations and Machine Learning Approaches

Bahadir Rusen Argun, Antonia Statt

Modern synthesis methods have enabled the fabrication of colloids with various different shapes. In addition, plastic waste degrades to nano- and microplastics particles in the environment, also leading to irregular shaped colloidal particles. These particles, polluting our habitats, have unknown implications for our health and environment. Therefore, understanding self-assembly of these non-spherical particles is crucial not only for many applications, like bottom-up fabrication of large ordered systems, but also for understanding the fate of microplastic particles in the aquatic environment.

To simulate the non-spherical colloids, we utilized the commonly used composite bead approach to represent the different shapes. We find that this approach is not only expensive as the complexity of the colloidal shape increases, due to inter-body distance calculations, but also cumbersome as it is not always obvious how the composite beads should interact with each other. Here, we will highlight the use of data-driven methods to accelerate our simulations. We trained neural networks to directly predict energy, or. forces and torques between non-spherical rigid particles based on their configuration, bypassing the need for inter-bead distance calculations. We show that MD simulations performed with neural net predicted forces and torques can accurately reproduce the structure and kinetics of the traditional simulations with explicit distance calculations. This approach can yield significant computational speedup and can be applied to irregular shapes with any pair interaction, given sufficient training data.

11e A Flow-Matching Approach for Generative Backmapping of Biomolecules

Michael Jones, Andrew Ferguson

Backmapping is the process of recovering all-atom detail from a coarse-grained simulation. In addition to generating physically plausible structures, a robust backmapping approach should produce an ensemble of structures that reflects the diversity of the one-to-many transformation. We leverage the recently developed flow-matching paradigm to learn a vector field that transforms a prior distribution representing coarse-grained structures to a data distribution representing all-atom structures. We build our prior by mapping atoms to their nearest coarse-grained bead and normally distributing their positions around that bead. We then train a time-conditioned equivariant graph neural network on the linear interpolation between the prior and ground-truth structures. Finally, we then leverage the learned vector field to integrate an ordinary differential equation that transforms out-of-sample coarse-grained samples into corresponding all-atom ensembles. We demonstrate our model on both protein trajectories and on DNA-protein trajectories using a 3-site-per-nucleotide coarse-grained mapping. Our results suggest a backmapping paradigm that is both generalizable across classes of biomolecules as well as across coarse-grained models.

11g A Database of Chemical Reaction Templates for REACTER

Jacob Gissinger

REACTER is a method for modeling chemical reactions in classical molecular dynamics (MD) simulations. We introduce a comprehensive library of reactions compatible with classical force fields and relevant to organic and polymer chemistry, which transforms REACTER into a plug-and-play reactive MD tool. The reaction library is enabled by the type label framework recently introduced into LAMMPS by the author, allowing database entries to be expressed in a format that is portable between different simulations. Previously, users were required to create reaction templates from scratch for every new reaction mechanism they wanted to model. A library for REACTER allows a particular reaction to be specified by name or queried via a substructure search, and the reaction database is referenced to activate that mechanism in the simulation. Automating this step in the workflow eliminates the most time-consuming and error-prone part of using REACTER, greatly streamlining the process of adding chemical reactivity into classical force fields. In particular, scanning through complex configurations of various polymer chemistries is made possible. This automated generation of atomic configurations for polymeric and composite materials is a particularly challenging bottleneck in the polymer informatics field. We demonstrate a workflow for incorporating the resulting properties of each configuration into optimization or machine learning schemes.

11h Webff: A Force-Field Data Repository for Classical Molecular Dynamics (MD) Simulations at Multiple Levels of Granularity

Frederick Phelan Jr.

This talk will describe WebFF, an open and extensible web-based force-field (FF) repository designed to support classical molecular dynamics (MD) calculations for organic compounds and related soft materials. The repository is built using the NIST Configurable Data Curation System (CDCS). The CDCS features a Django web-based interface, a Representational State Transfer (REST) API and a MongoDB database to support ontology based data descriptions using XML schema [1]. The initial release of the repository features four integrated XML schemas which describe different classes of force-field data. The first two are for Class I and Class II atomistic force-fields, respectively, the third accommodates coarse-grained force-fields, and the fourth encompasses data for mesoscopic models [2]. Users interact with the repository through two main portals. The Data Curation Portal supports upload of published force-field data with appropriate metadata descriptors to support provenance based data sharing. The User Portal supports search for curated force-field data based on the metadata descriptors and download in a number for common formats. New data entered into the repository can be assigned a persistent identifier in order to facilitate direct access, and the metadata descriptors enable keyword search and reference links to published articles or other websites to provide data provenance. The datasets may also include attachments. We will partner with developers to develop new schemas, curate data, develop custom exporters to support specific applications and expand force-field assignment tools. We also envision partnering with groups to stand-up other instances which can be interconnected (via the CDCS infrastructure) for federated searches and data sharing. User search, download and data curation to the site will be described and demonstrated.

References

[1] Configurable Data Curation System (CDCS), ULR=https://www.nist.gov/programs-projects/configurable-data-curation-system-cdcs

[2] WebFF-Documentation, URL=https://github.com/usnistgov/WebFF-Documentation

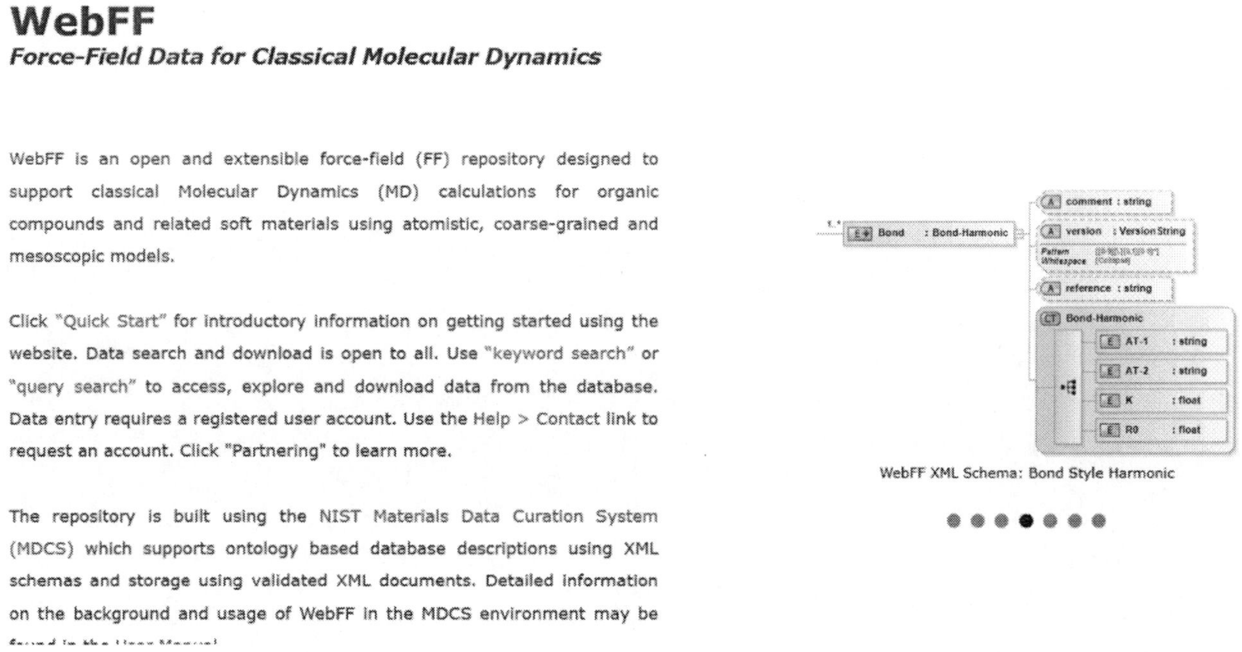

11i Benchmarking Molecular Simulations Using Formal Proofs

Tyler R. Josephson, Ejike Ugwuanyi, John Velkey, Colin Jones

When developing new methods for molecular simulation, eliminating bugs can be challenging. Programming languages like Python, FORTRAN, and Julia flag syntax errors, but don't (and cannot) catch errors in math or program logic - these must be rooted out by human experts. These issues can be managed by following best practices in software development (such as writing tests alongside the program), but even these do not guarantee that code is correct.

Lean is a new programming language developed for writing and automatically checking the logic of math proofs. We can translate derivations in science and engineering into math proofs, and use Lean's capabilities as a theorem prover to verify that the logic of the derivation is correct [1]. Lean also can be used to write executable programs, though this requires adopting a functional programming style, replacing for and while loops with recursive functions. These functions can be the subject of proofs that verify their properties. We use this approach to develop energy calculations for Lennard-Jones particles in periodic boundaries, and compare our implementation to the NIST benchmark [2]. The Lean implementation is a stronger benchmark than that available by NIST because the underlying mathematics have been formally verified.

[1] Bobbin, M.P., et al., "Formalizing Chemical Physics Using the Lean Theorem Prover," Digital Discovery, 3, 16 Nov 2024, p 264–280, https://doi.org/10.1039/D3DD00077J.

[2] Shen, V.K., Siderius, D.W., Krekelberg, W.P., and Hatch, H.W., Eds., NIST Standard Reference Simulation Website, http://doi.org/10.18434/T4M88Q.

11j Unified Multiscale Modeling for Catalysis: From Atomistic Insights to Reactor Design

Nestor Aguirre

In recent decades, complex screening studies have emerged to better understand structure-activity relationships for a wide array of materials and molecules. Not only can our multiscale modeling strategy optimize reaction processes, but it can also accelerate the discovery of novel catalysts.

Under the framework of the EU-funded ReaxPro project (https://www.reaxpro.eu), we have developed a multiscale modeling workflow that combines the versatility of the Amsterdam Modeling Suite (AMS) for atomic-scale simulations (including density functional theory (DFT) and machine learning potentials), automated energy landscape explorations with EON (https://theory.cm.utexas.edu/eon/), and Zacros (https://zacros.org) for dynamic modeling of heterogeneous catalysis. This seamless workflow is enhanced by an intuitive graphical interface and automatic search methodologies for transition states at different levels of theory. In particular, our recent integration of Quantum Espresso (https://www.quantum-espresso.org) with AMS allows us to leverage its advanced DFT capabilities, enabling highly accurate energy barrier calculations.

Our strategy enables the exploration of vast chemical spaces using accurate atomistic simulations. We aim to create a user-friendly platform that provides a comprehensive multiscale modeling framework, from atomistic level to continuum modeling of catalytic materials and reactors. This workflow has the potential to revolutionize the catalysis industry by significantly reducing the time and cost of discovering optimal conditions and materials. We will present illustrative examples of different catalytic processes and the current status of integration with the Computational Fluid Dynamics code CatalyticFoam (https://www.catalyticfoam.polimi.it) to bridge the gap to reactor-scale modeling.

11k Ab Initio Molecular Dynamics with Quantum Hardware and Transfer Learning

Clay Batton, Abid Khan, Prateek Vaish, Yaoqi Pang, Michael S. Chen, Grant M. Rotskoff, J. Wayne Mullinax, Brenda Rubenstein, Norm Tubman

The ability to perform ab initio molecular dynamics with both high accuracy and long timescales would allow for the precise determination of thermodynamic and kinetic properties in such processes as catalysis and biophysics. Here, we present a general framework to do so utilizing neural network potentials and quantum hardware. Fitting neural network potentials to energetics and forces obtained through quantum hardware would allow for such highly accurate and fast simulations. However, this process is limited by the current noisy energies computed on quantum hardware, along with the difficulty in scaling to large systems.

To alleviate the current difficulties in quantum hardware we utilize transfer learning, in which a neural network potential is trained initially on lower-quality data produced using classical computers and then fine-tuned on high-quality data generated on quantum hardware. Here, an initial neural network potential is fit to data from density functional theory, which is used to probe the behavior in the system of interest. To generate the configurations in the quantum hardware dataset, we use an active learning procedure to identify informative configurations for the subsequent refining of the neural network potentials. High accuracy results for this dataset are then obtained using the variational quantum eigensolver. This approach significantly reduces the quantum training dataset size while maintaining high accuracy of the potentials, allowing us to obtain accurate thermodynamic and kinetic information about model reactive systems.

11l Use Matlab to Compute Chemical Interactions on Quantum Computers

Hossein Jooya, Sudeep Ghosh, Sohini Sarkar, Aycan Hacioglu

In this presentation, we are going to showcase the mapping and computation of the chemical interaction Hamiltonian and its time propagation using Quantum Computing in MATLAB. We will introduce and give a brief overview of the MATLAB Quantum Computing Support Package. We will walk through a couple of demos showing how to build, simulate, and run quantum algorithms in MATLAB. Using the support package, you can prototype algorithms to accelerate solving problems such as optimization, scenario simulation, and machine learning, as well as currently unsolvable problems in chemistry.

516a Far and Few in between: Probing Structure-Selectivity Relationship in Membranes Using Path Sampling Techniques

Amir Haji-Akbari, Brian Shoemaker, Hessam Malmir, Tiago Domingues, Omar Khalifa

The ability of semipermeable membranes to selectively impede the transport of undesirable ions and molecules is key to many applications, from desalination and gas separation to biological membrane transport. For instance, membranes that are only permeable to water and reject most other ions and molecules are used in water desalination. Designing more selective membranes requires characterizing the kinetics and mechanism of the transport of the species rejected by the membrane. The timescales associated with such processes, however, can be too long to be accessible to conventional MD simulations. Moreover, the driven nature of the underlying separation processes excludes the utilization of advanced sampling techniques that require microscopic reversibility. Finally, the separation of timescales between the transport of desirable species (such as water) and undesirable species (such as salts) will result in considerable changes in reservoir concentration throughout an MD trajectory, which can skew the kinetics and mechanism of transport in ways that are difficult to quantify and correct.

Recently, we utilized [1] jumpy forward flux sampling (jFFS) [2], a path sampling technique developed in my group, and non-equilibrium MD to study pressure-driven ion transport through nanoporous graphitic membranes. Our approach addresses all these challenges. It not only allows us to accurately and efficiently estimate arbitrarily long mean first passage times in pressure- and field-driven filtration processes, but also to compute fluxes and permeabilities within a pseudo-equilibrium ensemble in which both reservoirs are at different- but almost constant- chemical potentials.

In this presentation, I will discuss several technical aspects of this new approach, particularly rigorous models based on classical electrostatics that we have developed to correct for the rather strong finite size artifacts in simulations of non-equilibrium ion transport [3-4]. What is remarkable about these models is that they rectify finite size effects "on the spot", i.e., from the information obtained from a single finite simulation. Therefore, the behavior of the system in the thermodynamic limit can be inferred without a need to conduct multiple simulations of systems with different sizes (as is the common practice in the computational chemistry community). I will also discuss some of our more recent findings on how pore arrangement within a membrane can impact transport resistance and selectivity in nontrivial ways [5]. In particular, we demonstrate that pores decorated with dipoles can exhibit strong correlations at the nanoscale in a manner that detrimentally impacts solvent-ion selectivity.

[1] Malmir H, Epsztein R, Elimelech M, Haji-Akbari A, *Induced charge anisotropy: a hidden variable affecting ion transport through membranes*, **Matter**, 2: 735 (2020).

[2] Haji-Akbari A, *Forward-flux sampling with jumpy order parameters*, **J. Chem. Phys.**, 149: 072303 (2018).

[3] Shoemaker BA, Domingues TS, Haji-Akbari A, *Ideal conductor model: An analytical finite-size correction for nonequilibrium molecular dynamics simulations of ion transport through nanoporous membranes*, **J. Chem. Theory Comput.**, 18: 7142 (2022).

[4] Shoemaker BA, Haji-Akbari A, *Ideal conductor/dielectric model (ICDM): A generalized technique to correct for finite-size effects in molecular simulations of hindered ion transport*, **J.**

Chem. Phys., 160: 024116 (2024).

[5] Shoemaker BA, Khalifa O, Haji-Akbari A, *Correlations in charged multipore systems: Implications for enhancing selectivity and permeability in nanoporous membranes*, **ACS Nano**, 18: 1420-1431 (2024).

516b Physics-Inspired Machine Learning Potentials for Unraveling Mysteries in Electrolyte Solutions

Shuwen Yue

Understanding the interactions which governs aqueous electrolyte solution properties is essential to advancements in countless domains, from electrochemistry to biochemistry. While molecular simulations have been extensively applied to study these systems, conventional empirical force fields fail to accurately predict many thermodynamic and dynamic properties, indicating limitations in capturing the underlying physics. In this talk, I will present our efforts in developing machine-learning potentials (MLPs) fit to a first-principles derived potential energy surface that can overcome these deficiencies. Moreover, we decompose energetics to identify physical underpinnings that contribute most to the property predictions. I will also describe our efforts in applying these potentials to better understand how electrolytes behave in confined environments and at hydrophobic interfaces. These studies highlight applications of machine learning for molecular simulations to not only improve property predictions but also achieve greater physical interpretability.

516c Understanding Ionomer Membrane Structure Via Analysis of Small Angle Scattering Profiles and Coarse-Grained Molecular Simulations

Jason Madinya, Stephen Kronenberger, Arthi Jayaraman

Ionomer membranes are used as proton conducting membranes for fuel cell and water electrolyzer applications due to their chemical stability, ion conductivity, and mechanical rigidity. The ion transport and mechanical properties of these membranes are dictated by the morphology of the hydrated ionomer film. The hydrophilic domains of hydrated ionomers are of particular interest as the structure and connectivity of these domains influence the ion-transport capabilities of the membrane. It is expected that the hydrophilic domain connectivity, shapes, and sizes will change with hydration, and that material processing can introduce nanoscale structural anisotropy. However, the structure of these domains is still not fully understood. In this work, we study the structure of hydrophilic domains with varying ionomer design and hydration using bottom-up and top-down approaches. In the bottom-up approach, we use coarse-grained molecular dynamics simulations with Martini_v2 model to show how real-space nanoscale structural features and multi-scale scattering profiles of simulated hydrophilic domains change with varying ionomer design.

In the top-down approach, we analyze data from small angle X-ray scattering (SAXS) measurements to understand processing-induced structural anisotropy of hydrophilic domains. Analysis of 2D scattering data - intensity I vs. magnitude of wave vector q and azimuthal angle \theta – from SAXS measurements of materials with anisotropic structures is not easy with conventional analytical models that usually focus fully (azimuthally) averaged 1D profiles. To analyze 2D scattering profiles, we have developed a new approach using random field structure reconstruction (RFS). Using RFS, we analyze the 2D SAXS measurements of as-cast or processed materials and identify real-space structural features (e.g., distribution of domain sizes, shapes, and orientational order) and 3D multi-phase structures. Having this type of structural understanding within ionomer membranes further enables calculations of proton transport for various ionomer designs.

516d Assembly of Short Amphiphilic Peptoids into Nanohelices with Controllable Supramolecular Chirality: Insights from Computational Simulation and Theoretical Analysis

Renyu Zheng, Mingfei Zhao, Jingshan Du, Tarunya Sudarshan, Yicheng Zhou, Anant Paravastu, James J. De Yoreo, Andrew Ferguson, Chunlong Chen

Computational modeling and theoretical analysis are pivotal for gaining a deep understanding of the structures and functions essential for the rational design of synthetic biomaterials that mimic natural counterparts. In this work, we present the formation of a series of nanohelices derived from a well-established class of peptide-mimetics known as peptoids (poly-N-substituted glycines). These peptoid nanohelices are designed to mimic the helical protein assemblies crucial in biological systems. By integrating experimental mechanistic studies, molecular dynamics simulations, and theoretical analysis, we demonstrate that the structures of nanohelices and their supramolecular chirality can be precisely controlled through side-chain chemistry. Molecular dynamics simulations unveil the thermodynamically favorable ordered packing of these amphiphilic peptoids and the transition process from flat sheets to twisted nanofibrils. Furthermore, we have developed a phenomenological free energy model to elucidate the self-assembly of the twisted ribbons. This model provides insights into the twisted ribbon geometry as a function of solvent conditions and intermolecular attractions. Our research indicates that amphiphilic peptoids minimize the exposure of hydrophobic domains by adopting a twisted helical conformation, underscoring the significant roles of both polar and hydrophobic domains in nanohelix formation. The ability to precisely control the structural organization of sequence-defined peptoids and their supramolecular chirality of their assemblies opens the door to a system suitable to reveal how molecular interactions between fundamental chemical moieties control the arrangement, packing, and assembly of biomacromolecules and inorganic particles, generating a wide range of protein-like, functional nanomaterials as nanocarriers for drug delivery, as artificial enzymes for catalytic reactions, and as scaffolds for biomimetic mineralization. Overall, our findings establish a foundation for designing and synthesizing chiral functional materials using sequence-defined synthetic polymers.

516e Dynamics and Drying-Induced Assembly of Shape-Anisotropic Nanoparticles

Michael Howard

Supraparticles assembled from nanoparticles are promising materials for catalysis, gas adsorption, drug encapsulation, and photonics. A supraparticle's properties and function are strongly connected to its internal structure. This structure is known to depend on the nanoparticles and how the supraparticle was fabricated, e.g., by casting a droplet and removing the solvent at a certain rate. There is particular interest to use nanoparticles with anisotropic shapes (such as rods) to create supraparticles with controlled porosity or optical properties, but there are significant gaps in our understanding of, and hence ability to engineer, this dynamic self-assembly process. Here, we first use mesoscale particle-based simulations (multiparticle collision dynamics method) to study the dynamics of various shape-anisotropic nanoparticles (octahedra, cubes, tetrahedra, and spherocylinders) in bulk suspensions. We then use simulations in close collaboration with experiments to investigate the properties of supraparticles self-assembled from either rod-like nanoparticles or a mixture of rod-like and spherical nanoparticles. Our work sheds important insight into the dynamics of shape-anisotropic nanoparticles and suggests strategies for tuning the structure, and hence properties, of supraparticles made from them.

516g Developing Data-Driven Methods to Enable Computationally Efficient First-Principles Kinetic Models in Nanoporous Catalysts

Brandon C. Bukowski

Crystalline nanoporous solids such as zeolites are employed in a wide range of chemical processes due in part to their tunable micro-environments. Kinetics at internal zeolite acid sites can be modified by changing pore size, pore architecture, or polarity.[1,2] These environments impart shape-selectivity that preferentially stabilizes transition states, but the large design space including pore architecture, polarity, and catalytic active site identity precludes comprehensive kinetic studies. Density functional theory (DFT) describes the electronic states of reactive intermediates and transition states but cannot access the longer length scales necessary to quantify adsorbate entropy and the fluxionality of coadsorbed species or solvents. Force field-based simulations can accurately capture these conformational changes but require parameterized values. Machine learning interatomic potentials have emerged as a technique to derive parameterized models from DFT data, and thereby link DFT and force field-based observables.

We used zeolite-catalyzed lactic acid dehydration to acrylic acid[3] and ethanol-to-olefins[4] as probe reactions to interrogate the relative importance of intrinsic kinetic barriers and mass transfer on product selectivity. We then developed machine learning strategies to accelerate the quantification of hopping rate constants and quasi-harmonic adsorbate entropies under reaction conditions and demonstrated their relevance for predicting turnover frequencies, reaction orders, and product selectivities using microkinetic models. These methods are scalable for larger reaction pathways allowing for the development of ab-initio microkinetic models that account for quasi-harmonic interactions and improved predictions of reaction kinetics.

1. Bukowski, B. C. et al. "Defect-Mediated Ordering of Condensed Water Structures in Microporous Zeolites". *Chem. Int. Ed.* 2019.
2. Bregante, D. T. et al. "Cooperative Effects between Hydrophilic Pores and Solvents: Catalytic Consequences of Hydrogen Bonding on Alkene Epoxidation in Zeolites". *Am. Chem. Soc.* 2019.
3. Yan, B. et al. "Acrylic Acid Production by Gas-Phase Dehydration of Lactic Acid over K+-Exchanged ZSM-5: Reaction Variable Effects, Kinetics, and New Evidence for Cooperative Acid–Base Bifunctional Catalysis". *Eng. Chem. Res.* 2020.
4. Zhang, J. et al. "Isolated Metal Sites in Cu–Zn–Y/Beta for Direct and Selective Butene-Rich C3+ Olefin Formation from Ethanol". *ACS Catal.* 2021.

54a Accelerating Polymer Design with Targeted Properties Using Machine Learning and Physics-Based Models

Alex Chew, Mohammad Atif Faiz Afzal, Anand Chandrasekaran, Jan Henk Kamps, Ramakrishnan Vaidyanathan

Designing new, industrially relevant polymers is challenging because of the need to optimize multiple materials' properties simultaneously, which is expensive and often infeasible using traditional trial-and-error approaches. One possible solution to identifying promising polymeric materials is to employ a combination of machine learning and physics-based tools to screen the polymer design space and provide suggestions for new polymers that meet the criteria for an industrial application. In this work, we demonstrate a workflow that utilizes machine learning and molecular modeling approaches to design new polymers (specifically, polycarbonates) that satisfy five polymer properties, including the glass transition temperature, optical properties, and mechanical properties. Using a relatively modest dataset of fewer than 200 points, we developed quantitative structure-property relationship (QSPR) models to accurately predict the experimental polymer properties given the homo- or co-polymer structures and composition as input. Leveraging these computationally efficient QSPR models, we then screened over ~10,000 polymer structures that were generated through R-group enumeration tools. We used these QSPR predictions to create multi-parameter optimization scores to help down-select the large polymer space to ~10 promising candidates. We validated the predicted properties of the top polymer candidates using classical molecular dynamics simulations and density functional theory, which revealed reliable correlation between physics-based and QSPR approaches. Finally, we validated the computational predictions against experiments, which showed good agreement with QSPR and physics-based models. Our workflow demonstrates the usefulness of combining data-driven and physics-based approaches in designing new polymers given a small dataset, which is broadly useful for scientists interested in leveraging computer-aided strategies to innovate new materials while mitigating the need for extensive trial-and-error experimentation.

54b Modeling LLDPE Resins in Multilayer Film Heat-Sealing with Slip-Link Model

Marat Andreev, Jeffrey D. Weinhold, Jonathan Moore, Gregory Rutledge

Optimization of the industrial processing of polymers requires accurate modeling of the rheological behavior. There are a number of molecular models for the rheology of polymeric materials. Their main advantages include a direct connection between molecular level details of polymer chains and macroscopic behavior of the material. However, such models are challenging to implement due to computational cost requirements for realistic molecular weights. Generally, significant additional effort is required to develop, optimize and apply these models to industrially relevant cases.

Heat sealing is a common industrial method of hermetically sealing flexible packaging made from polymer films. The typical heat-sealing operation involves hot bars that bring two polymer films together and apply pressure and heat to form the seal. The polymer films typically comprise multiple layers; different layers serve different purposes and contain different polymers. The simplest multilayer structure includes a toughness layer and a sealant layer. The sealant layer is designed to melt, flow and recrystallize during the operation cycle.

Here, we take the discrete slip-link model, an established computational rheological model for entangled polymers, and apply it to the simulation of heat sealing of multilayer LLDPE films. The model is attractive due to its applicability to broadly polydisperse polymers under non-linear flows. Additionally, we recently extended the model to semicrystalline and crystallizing polymers. We describe our efforts to adapt the model for heat-sealing simulations. The modifications include optimizing the computational cost and inverting the model's stress-strain relation. Next, we combine the rheological model with the heat transfer and crystallization kinetics models. The combined model of multilayer film sealing features toughness and sealant layers. A layer representation in the model is based on inputted molecular and rheological characterization information for each LLDPE resin. The model calculates temperature, crystallinity profiles, stresses, flow rates, film thickness evolution, and flow volume, from which conclusions can be drawn about hermeticity of the seal.

54c Toward Cryogenic Membrane Gas Separations: Insights from Molecular Simulation and Statistical Mechanics

Cristel Carolina Brindis Flores, Walter Chapman

Membrane gas separations at cryogenic temperatures rely on the relative permeability of gaseous components through polymeric materials. Although gas separations have been widely studied at ambient temperatures, there is limited data and simulation results at cryogenic conditions. With limited experimental data, molecular simulation offers a powerful tool to analyze the transport and thermodynamic properties of these gases within polymers. Additionally, advanced equations of state, such as PC-SAFT, can accurately predict thermodynamic properties across a wide temperature range.

In this study, we embark on a systematic study of the diffusivity and solubility of various gases within polymer films using molecular dynamics simulations. To complement these simulations, we employ an extended version of Widom's Potential Distribution Theory (PDT) to determine the chemical potential. Additionally, we will use regularization method, known as QCT-PDT, to evaluate the work required to create cavities within the material.

By combining diffusivity and chemical potential calculations, we aim to derive the necessary data for permeability assessments across polymer films. Through these calculations, we can estimate the exchange of gaseous species within the film. Our methodology will be validated against experimental data, and we will demonstrate the efficacy of PC-SAFT in fitting molecular simulation results for the chemical potential of gaseous components in the polymer. This fitting will enable us to interpolate and extrapolate molecular simulation findings effectively.

Overall, this research promises to shed light on the intricate dynamics of gas transport within polymer films at cryogenic temperatures, offering valuable insights for various industrial applications.

54d Exploring Gas Diffusion in Carbon Monoxide Dehydrogenase/Acetyl-CoA Synthase (CODH/ACS) Enzyme Complex Using Molecular Simulations.

Suman Samantray, Bojana Ginovska-Pangovska, Simone Raugei

The carbon monoxide dehydrogenase (CODH) and acetyl-CoA synthase (ACS) enzymatic complex catalyzes the (reversible) electrochemical conversion of CO_2 to CO and the subsequent condensation of CO with a methyl intermediate and coenzyme A to form acetyl coenzyme A. CO_2 reduction happens at the CODH C-cluster, and the CO product is transported and consumed at the ACS A-cluster. The bifunctionality of the complex is facilitated by a 140 Å-long gas channel for the transport of CO between the C- and A-clusters.

CODH and the CODH/ACS are central for atmospheric carbon monoxide oxidation and supporting microbial life. These enzymes are also the focus of recent biotechnological applications because of their potential replacement of two energy-intensive industrial processes: 1) the water–gas shift reaction, a widely used industrial process for H_2 or CO production, and 2) the Monsanto process used to produce acetic acid from methanol and CO. However, the effective and widespread use of CODH and CODH/ACS for industrial fields is limited by their poor activity under aerobic conditions. One strategy to design O_2-tolerant complexes is to prevent O_2 from entering the enzyme matrix.

In this work, we use atomistic free energy simulations to explore the gas diffusion through the CODH/ACS complex from M. thermoacetica and other design variants that favor uptake of CO_2 and O_2. We also explore how the transport mechanism is affected by conformational changes in the subunits of the enzyme caused due to transformation of cavities and gating mechanism of residues.

Research Interests:

"Peptoids" as a viable biomaterial in food packaging industry.

Antifreeze proteins (AFPs) stand as a paradigm-shifting innovation, that prevent freezing in organisms living in sub-zero environments by inhibiting ice crystallization. AFPs, such as Type I AFP from winter flounder and Type III AFP from ocean pout, have been instrumental in preventing ice crystallization in biological and industrial settings. This unique function has profound implications in cryopreservation, organ transplant technology, and notably in the food industry, where AFPs play a pivotal role in enhancing product longevity and quality.

Peptoids are synthetically engineered polymers, that could mimic natural AFPs with enhanced flexibility. The fundamental distinction between peptoids and peptides, while structurally similar, differ in the placement of their side chains. In peptides, side chains are attached to the alpha carbon atoms, whereas in peptoids, they are bonded to the nitrogen atoms of peptide backbone. Peptoids could be customized with targeted functional properties, to increase stability against enzymatic degradation, higher resilience under various physical conditions, and a reduced risk of immune responses, making them more suitable for industrial applications.

The research will address critical scientific questions: How do structural variations in peptoids influence their binding efficiency and antifreeze activity compared to natural AFPs? What thermodynamic and kinetic parameters are crucial in determining the antifreeze efficiency of peptoids? How can molecular dynamics simulations contribute to the design of optimized peptoid AFPs for industrial applications, particularly in food packaging?

Circular economy: Conversion of biomass-plastic waste to biochar-based adsorbents.

The crisis of plastic waste, particularly the accumulation of microplastics in the environment, presents a critical challenge for sustainable development. Addressing this issue requires innovative solutions that not only decompose plastic waste but also repurpose it into beneficial materials. Artificial metalloenzymes (ArMs) offer a targeted approach to catalyze the degradation of resistant plastic polymers, transforming them into value-added products such as biochar. Biochar, derived from the carbonization of organic matter, is recognized for its adsorptive properties, and is proposed mechanism to sequester pollutants.

Natural metalloenzymes like laccase and manganese peroxidase catalyze essential reactions in biological systems, including the degradation of complex molecules. Mimicking natural enzymes, artificial metalloenzymes can be engineered to target the sturdy chemical bonds in plastics such as polyethylene terephthalate (PET), high-density polyethylene (HDPE), and polystyrene (PS). Advanced molecular simulation techniques coupled with rational design and directed evolution can finely tune the active sites of metalloenzymes for specific substrate interaction, enhance catalytic rates, and provide stability against harsh industrial conditions.

The research aims to: Design artificial metalloenzymes with enhanced specificity for breaking down multiple types of plastics, focusing on the creation of intermediates destined for biochar conversion. Explore metabolic engineering of microorganisms to host these ArMs, evaluate the life-cycle of the enzymatic byproducts, and upscale the entire process for industrial use. Successful implementation would represent a dual achievement: mitigating plastic pollution and harnessing the resulting biochar for environmental adsorption applications, thereby fostering a circular economy in waste management.

PFAS-mediated dysregulation of biomolecular condensates.

The pervasive presence of per- and polyfluoroalkyl substances (PFAS) in the environment, coupled with their long biological half-life, has emerged as a concern for public health. PFAS compounds like PFOA and PFOS, notorious for their toxicity, exposure to it is linked to a multitude of health risks, and there is a critical need to understand their interaction with cellular components at a molecular level. This research focuses on the impact of PFAS on biomolecular condensates, which are crucial for various cellular processes, including RNA metabolism, signal transduction, and the stress response.

Biomolecular condensates, formed via liquid-liquid phase separation, are highly dynamic and sensitive to changes in their microenvironment. The physicochemical properties of PFAS, such as their hydrophobicity and stability, make them potential disruptors of the condensates' structural integrity. PFAS affect biomolecular condensates at the molecular level, this research aims to utilize molecular simulations to bridge the gap between environmental exposure and cellular dysfunction, offering a foundational perspective on the health implications of PFAS. This work has the potential to inform risk assessments, shape regulatory decisions, and inspire the development of therapeutic interventions to counteract the detrimental effects of PFAS on human health.

Teaching Interests:

The beauty of academic life is that it gives an opportunity to learn, teach, and work with young, bright, and dynamic students, who are the future of our society. Through the influences of the great teachers, I am at this stage to pursue academic life as a part of devoted teaching and research to serve our society. In the following, I describe in detail experiences, teaching philosophy, and plan for teaching.

Teaching experience: I have had a privilege of being TA for the computational biotechnology practical courses of MSc students at University of Galway, Ireland and RWTH Aachen University, Germany. In addition, I co-mentored theses of BSc, MSc students, and students.

Teaching Philosophy: In my opinion, the most important aspect of teaching is that we learn, teach, work, and develop together for the welfare and service of our society. I believe that we can learn best by 'doing'. The fundamental aspect of teaching is igniting the mind of students with the curiosity and skills necessary to develop questions as well as guide them one-step further to motivate, promote and develop skills and scientific processes necessary to investigate the answers to those questions. To facilitate this, I will keep my classes interactive and open for discussion. Besides, I will take the opportunity to discuss teaching, learning, training, and research with students, friends, colleagues, senior professors, and experts for the betterment of my service to the institute and society.

Teaching Plan: I am interested in teaching the core and professional courses of Biochemical Engineering. In the core courses, I am in the teaching of Protein thermodynamics, Statistical mechanics, Chemical kinetics, and Chemical reaction Engineering. In the professional courses, I would like to teach Bioinformatics, Structural Biology, and Biophysics. Besides, I have the interest to design courses, e.g., Bioeconomy, Protein Engineering.

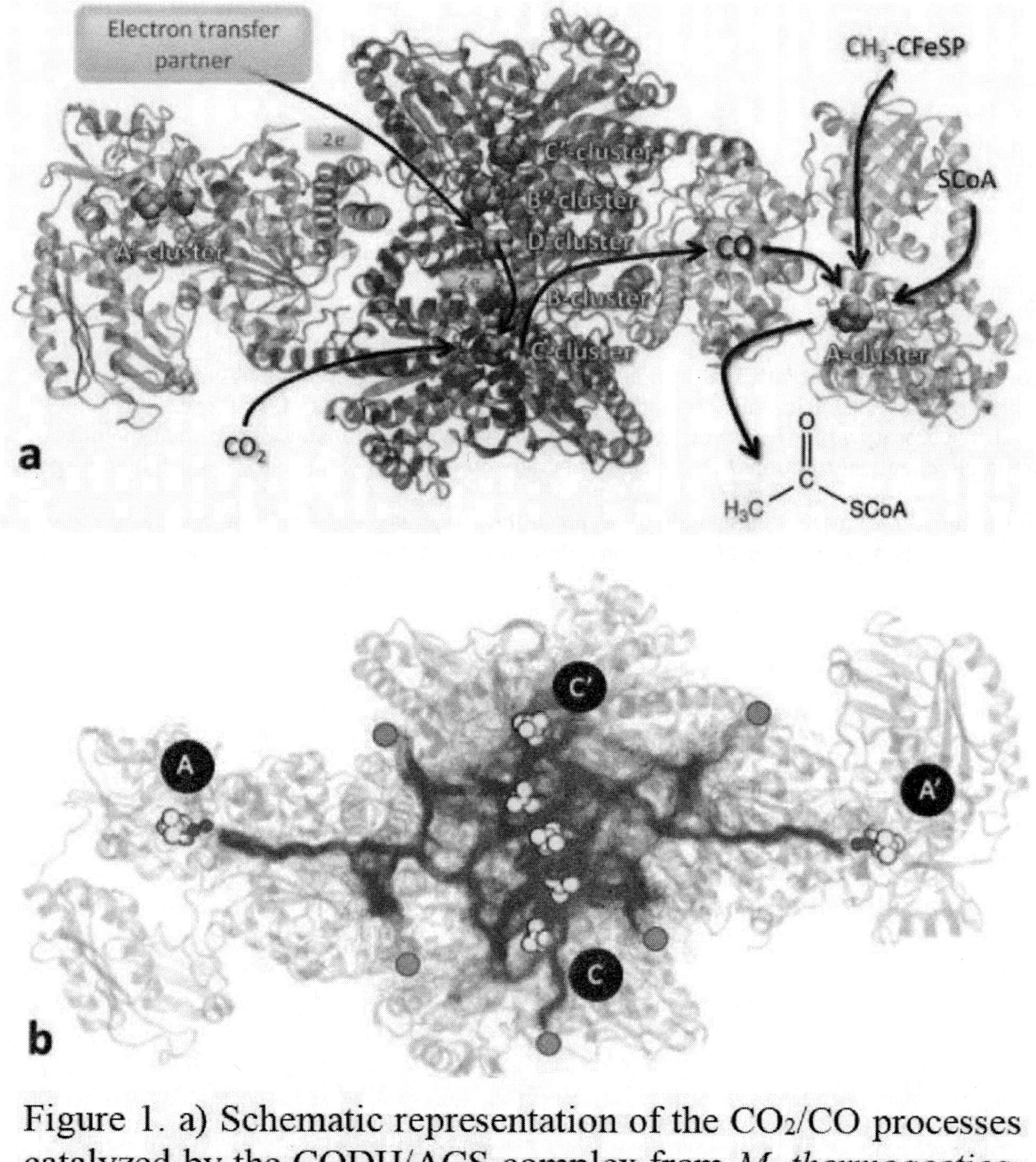

Figure 1. a) Schematic representation of the CO_2/CO processes catalyzed by the CODH/ACS complex from *M. thermoacetica*, with the active metalloclusters A- and C- labelled in cyan. b) A composite view of gas channels (dark gray) in CODH/ACS discovered by long-timescale molecular dynamics simulation.

54e Investigation of Pd-Based Alloy Membranes for Hydrogen Separation: An Application of a Universal Neural Network Potential

Taku Watanabe, Akihiro Nagoya

Pd-based alloy membranes for hydrogen separation are highly efficient and have several important practical advantages over traditional methods like pressure swing adsorption or cryogenic distillation. However, the use of Pd, a rare and expensive metal, and its reactivity with certain gas mixture components, pose challenges. To address these issues, Pd alloys have been widely explored in the literature.

In this work, we report our investigation of the surface composition and structures of the Pd-based alloys and their impact on the adsorption of gas species on the metallic membranes. Our unique proprietary universal neural network potential, called Preferred Potential (PFP), was applied to perform necessary atomistic simulations. One of the advantages of using PFP is the capability of simulating chemical reactions at near-DFT accuracy on much larger system or much higher speed.

We examined the surface segregation behavior of the alloys as a function of temperature and composition using Monte Carlo simulations. Pronounced segregation of the solute was found in fcc $Pd_3Ag(111)$ and $Pd_3Au(111)$, but solvent segregation was found in $Pd_3Pt(111)$. The segregation behavior is found to be significantly altered by the presence of adsorbate species and absorbed hydrogen in the bulk phase. These findings are consistent with the previously known facts in the literature.

MD simulations are also utilized to explore the dynamics of the gaseous species as well as the surface metal atoms. The MD simulations were found to allow structural relaxation in which the simple MC simulations cannot explore. The surface perturbation observed do vary depending on the composition of the alloy and the adsorbed species.

The outcome of this study will provide some important insights to the design of Pd-based alloy membranes and the method of evaluating their performance with atomistic detailed information.

54f Understanding Thermodynamics of the Cytochrome P450 Enzymatic Cycle for Ionic Liquid Biodegradation

Sudip Kumar Das, Jindal Shah

Ionic liquids (ILs) have been considered highly efficient solvents, co-solvents, or agents for applications in the chemical industry. They present numerous advantages over traditional solvents: low volatility, high thermal and chemical stability, tunable properties, etc. These unique properties enable them to be used in industrial processes such as ISOALKYL process for alkylation, mercury removal, battery electrolytes, lignocellulose dissolution, gas absorption, drug delivery, and organic catalysis, etc. Though ILs are termed as 'green solvents' their persistence in the environment could be a potential problem. Experimental investigations have shown that ionic liquids can be recalcitrant, partially biodegradable or fully biodegradable. The potential for biodegradation has been shown to be higher when the alkyl chain contains six or more carbon atoms. Various studies have hypothesized that iron porphyrin, an active site of bacterial cytochrome P450 protein, is responsible for the oxidative transformation of the alkyl side chain which initiates the ionic liquid biodegradation. The hydroxylation reaction by iron porphyrin occurs via a number of intermediate steps. This work presents a detailed thermodynamic analysis for each of the intermediate steps in the protein medium by adopting quantum mechanical calculations for 1-n-alkyl-3-methylimidazolium chloride (n = 2, 4, 6, 8, and 10) to understand the chain length-dependent biodegradation. Equilibrium geometries for each of the intermediate steps will be presented. Important geometrical criteria such as imidazolium ring position, distance between imidazolium ring and iron porphyrin ring, and cation-anion distances in protein medium for each intermediate steps will be discussed. Rate-limiting steps for ionic liquid hydroxylation reaction by iron porphyrin will be identified. This work also sheds light on the oxygen binding mechanism in the presence of ILs into the binding pocket. We will discuss the evolution of byproduct water to understand the regeneration of the initial state of the catalyst.

54g Unsupervised Clustering of Existing Simulation Studies on Asphaltene Aggregation

Felipe Perez, Azeezat Ali, Alberto Striolo

Asphaltenes are defined as a solubility class rather than a singular family of molecular compounds, therefore there is no unique molecular structure that corresponds to asphaltene. Perhaps as a consequence of this definition, direct probing of physical and chemical properties of asphaltene molecules has been historically challenging in the laboratory. It was only in the past few years that experimental imaging of asphaltene structures resolved representative molecular geometries. Could the identification deriving from these experiments be useful for moving forward our collective understanding of the properties of asphaltenes? Is the experimental identification of numerous asphaltene molecules implying that simulations should consider all these compounds for providing useful insights? Is it possible that the community has been considering a limited sample of asphaltene molecules when attempting to shed light on asphaltene behavior? This focused review attempts to address these questions with the help of unsupervised clustering. The results demonstrate that the experimental asphaltene structures can be clustered into four groups that resolve different physical properties of asphaltenes. Then we probe the existing literature to assess whether compounds representative of the molecules experimentally identified as 'asphaltenes' using non-contact atomic force microscopy (nc-AFM) (Schuler, B., et al., J. Am. Chem. Soc. 2015, 137, 9870; Schuler, B., et al., Energy Fuels 2017, 31, 6856) have been investigated via simulations. We compare structures used in computational studies found in the literature against the experimental geometries, and summarize their context and findings. The exercise reveals that only a few asphaltene molecular structures have been studied while several asphaltene structures have been overlooked in published simulation works. While this is due to our collective efforts to study a 'representative' structure for asphaltenes, our analysis reveals a bias in the systems that have been simulated, which leads to suggesting possible directions for future studies that might shed light on the mechanisms responsible for asphaltene precipitation.

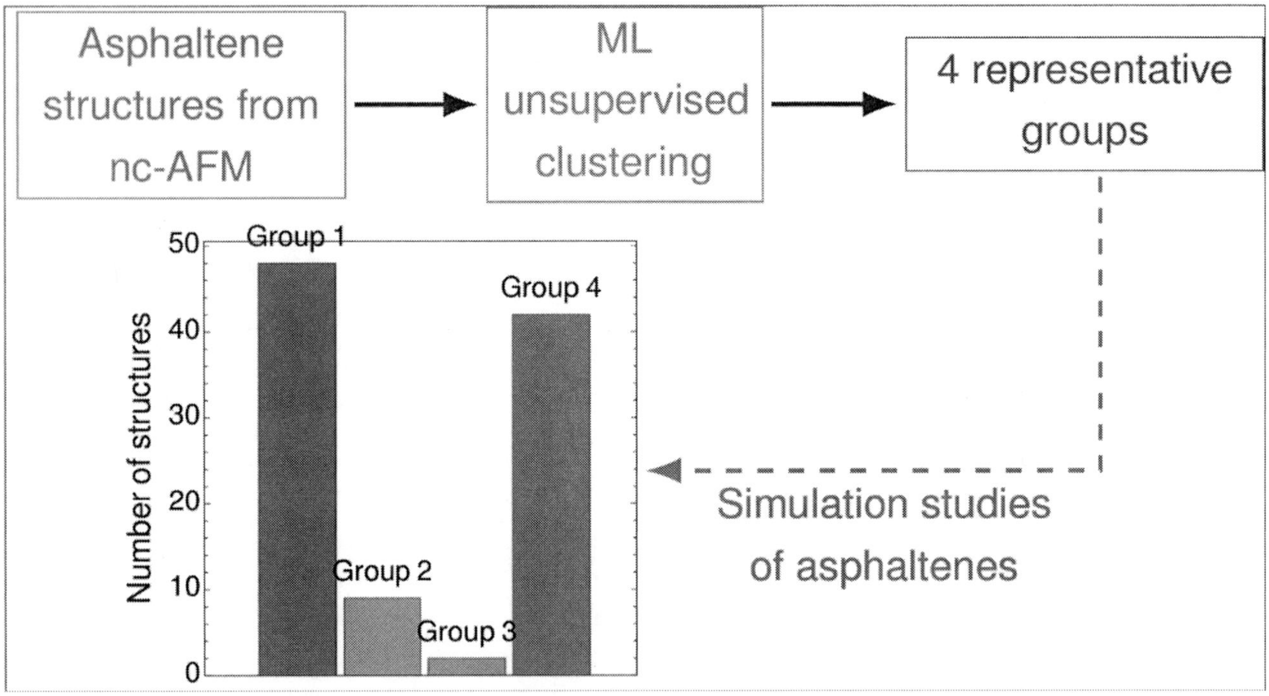

55a Understanding the Self-Assembly of Skin Lipids from Molecular Dynamics Simulations: A Multiscale Perspective

Clare McCabe

The barrier function of the human skin is primarily located in the stratum corneum (SC), which is composed of skin cells (corneocytes) surrounded by a dense, lamellar lipid matrix made up of ceramides (CERs), cholesterol (CHOL), and free fatty acids (FFAs) of various lengths. While the lipid structure of intact SC can be reproduced experimentally with model systems using synthetic lipids,[1] the organization and role of each lipid in determining the barrier properties of skin, is not well understood due to the complex nature of the SC lipid matrix. To address this need, atomistic molecular dynamics (MD) simulations of pre-assembled lipids in primarily bilayer configurations have been used to study the behavior of the SC lipids.[2,3] However, such simulations can suffer from initialization bias due to low lipid mobility. Furthermore, simple bilayers are not representative of the complex multilayer structure of the SC. To access the long timescales and large system sizes needed to self-assemble multilamellar structures, we have developed computationally efficient coarse-grained (CG) models using the Multi-State Iterative Boltzmann Inversion (MS-IBI) method[2] based on simulation data from fully atomistic models for commonly studied SC lipids.[3] The self-assembly of mixtures of CERs, CHOL, and FFAs are subsequently studied into multilayer assemblies. Atomistic configurations are then generated by back-mapping the self-assembled multilayer CG configurations and in turn provide insight into lipid organization and structure of SC lipid systems.

References

[1] J. A. Bouwstra and M. Ponec, "The Skin Barrier in Healthy and Diseased State," Biochimica et Biophysica Acta – Biomembranes, 1758, 2080-2095 (2006).

[2] P. Shamaprasad, C. O Frame, T. C. Moore, A. Yang, C. R. Iacovella, J. A. Bouwstra, A. L. Bunge and Clare McCabe, "Using Molecular Simulation to Understand the Skin Barrier" Progress in Lipid Research, 88, 101184 (2022).

[3] T.C. Moore, C. R. Iacovella, and Clare McCabe, "Derivation of Coarse-Grained Potentials via Multistate Iterative Boltzmann Inversion," J. Chem. Phys, 140(22) 224104 (2014).

55b Phase Transitions in Systems with Interconverting Species

Pablo M. Piaggi, Thomas J. Longo, Mikhail Anisimov, Betul Uralcan, Pablo Debenedetti

The separation of substances into different phases is ubiquitous in nature and important scientifically and technologically. This phenomenon may become drastically different if the species involved interconvert. In the presence of an external force large enough to overcome energetic differences between the interconvertible species (forced interconversion), the two alternative species will be present in equal amounts, and steady-state, restricted phase separation into mesoscales is observed. We investigate the formation of such structures through simulations of physically distinct microscopic models of binary mixtures that exhibit both equilibrium (natural) interconversion and a nonequilibrium source of forced interconversion. A chiral tetramer model whose enantiomers can interconvert exhibits rich fluid-phase behavior that may have implications for pre-biotic chiral symmetry breaking. The phase transition mechanism in this system does not involve nucleation. The model's critical point belongs to the 3D Ising universality class.

55c Stable Quasicrystal from Tetrahedron Colloids

Michael Engel

The tetrahedral geometry is ubiquitous in natural and synthetic systems. Regular tetrahedra do not tile space, which makes understanding their self-assembly behavior a formidable challenge. Simulations of hard tetrahedra –that is particles with the shape of a regular tetrahedron interacting only by excluded volume interactions– discovered a dodecagonal quasicrystal stabilized by entropy alone [1]. But while this quasicrystal forms robustly and reproducibly in simulation, it competes with periodic approximants and cannot be the thermodynamic ground state in the limit of infinite pressure. In this limit, the densest packing will eventually prevail, which is a simple (in comparison) dimer crystal [2]. After a brief historic review of early research in the Glotzer group on this topic, we advance research on tetrahedron particles in two directions. First, we discuss experimental realizations of phases of tetrahedron from gold nanocrystals where vertex sharpness, surface ligands, and the self-assembly environment play key roles in the formation of the quasicrystal, its approximants, and the dimer crystal [3]. Second, we discuss the growth of tetrahedron nanocrystals by spontaneous symmetry breaking of the cubic crystal lattice to tetrahedron Wulff shape symmetry during the crystallization process. We utilize Monte Carlo and molecular dynamics particles simulations as well as kinetic Monte Carlo modeling. Efficient code optimization and effective graphical visualization are key. As time permits, I will discuss other complex ordering phenomena that were initiated during my time in the Glotzer group and continued thereafter.

[1] A. Haji-Akbari, M. Engel, A.S. Keys, X. Zheng, R.G. Petschek, P. Palffy-Muhoray, S.C. Glotzer, Nature 462 (2009) 773.
[2] E.R. Chen, M. Engel, S.C. Glotzer, D. Comput. Geom. 44 (2010), 253.
[3] Y. Wang, J. Chen, R. Li, A. Götz, D. Drobek, T. Przybilla, S. Hübner, P. Pelz, L. Yang, B. Apeleo Zubiri, E. Spiecker, M. Engel, X. Ye, J. Am. Chem. Soc. 145, 17902 (2023).

55d A Careful Look at the Development of the Virial Equation of State

David A. Kofke

The virial equation of state (VEOS) has the distinction of being the only general-purpose thermodynamic model that has a direct, unambiguous connection to the intermolecular interactions, such that the coefficients of the VEOS may be computed rigorously from a given, arbitrarily complex molecular model. Practical application of the VEOS is limited however, due to failure of the series to converge for higher densities and condensed phases. It is interesting to consider the origins of this failure, and the manner in which the series represented by the grand-canonical partition function, which applies at all densities, is transformed into the VEOS, which does not. We identify a flaw in the textbook derivation of the series which, while not rendering the VEOS incorrect, does suggest how it can become inaccurate at high densities. Attempts to remedy this issue are described.

55e Self-Assembly Models to Simulate Phase Transformations

Julia Dshemuchadse

Self-assembly provides a path to create new materials, and it can lead us toward a better understanding of the fundamental building principles of matter—both hard and soft. With minimalistic models, we simulate the processes of crystal growth and solid–solid phase transformations to gain insight into mechanisms whose observation in experiments is particularly challenging. In this manner, we study complex crystal structures, the competition of ordering and disordering phenomena, and magic-size assemblies, with the ultimate goal to identify design targets and create new functional materials in the future.

55f Coarse-Grained Modeling of Crystal Growth Using Forcefields from Iterative Boltzmann Inversion and Particle-Swarm Optimization

Ronald Larson

Because of the infeasibility of directly simulating growth of drug crystals using molecular dynamics (MD) with atomistic force fields, we develop course-grained (CG) force fields (FFs) at two levels, with each CG bead representing either 5-7, or 2-3 "heavy" (i.e., non-hydrogen) atoms within ring-containing drug molecules. The first, highly coarse-grained, FF is produced using Iterative Boltzmann Inversion (IBI), while the second is obtained using Particle Swarm Optimization (PSO). Both are optimized to produce stable crystal unit cells for a representative drug molecule, carbamazepine, and a co-crystal of it with succinic acid. Both forcefields allow growth of crystal facets into a melt of the same molecules, but the more highly coarse-grained forcefield can give unrealistic structure of the melt, and in some cases unphysical growth rates of crystal facets, while growth rates of crystal facets derived for the model with less coarse graining are more realistic physically. In addition, the more resolved CG forcefield is better able to capture the effects of hydrogen bonding as well as conformational changes in the molecule, and are much better suited to simulation of co-crystals. Similar studies of a second crystallizing drug molecule, phenytoin, will also be presented. Various challenges encountered in this work, and the prospects for further work on coarse-graining of drug molecules for MD are discussed.

55g Thermodynamics and Kinetics of Ag Nanocrystals

Kristen Fichthorn

Metal nanocrystals are useful in a host of applications, ranging from catalysis to plasmonics to electronics to various personal and health applications, and many studies show there are optimal nanocrystal sizes and morphologies for these applications. In the interest of reliably synthesizing optimal nanocrystals, it is important to understand the shape trajectories nanocrystals follow as they grow. In a self-seeding synthesis, the small (likely single-nm) nanocrystal seeds that form after nucleation are fluxional and we expect them to assume an equilibrium shape distribution as they grow. As the nanocrystals grow larger, they may become locked into kinetic shapes. It is important to understand the thermodynamic-to-kinetic shape transition to design processing routes by which nanocrystals with specific sizes and shapes could be synthesized. The time scales over which shape transitions occur as nanocrystals grow are poorly understood at present and it is important to first understand the thermodynamics of nanocrystal shapes to make progress toward this goal.

I will recount our efforts to understand both equilibrium and kinetic shapes of Ag nanocrystals. To resolve equilibrium shapes, we use parallel tempering molecular dynamics (MD) simulations to probe the minimum free-energy shapes of Ag nanocrystals containing 100–200 atoms in a vacuum. Time permitting, I will discuss our simulations in a solution environment. Our simulations reveal a shape intermediate between a Dh and an Ih, a Dh-Ih, that has distinct structural signatures and magic sizes. At certain critical sizes, nanocrystal shapes can change dramatically with the addition and removal of a single atom or with a change in temperature at a fixed size. We use accelerated MD simulations (hyperdynamics) to study the deposition of Ag atoms onto nanocrystals at various temperatures and experimentally viable deposition rates. Our studies show that even the smallest (100-atom) nanocrystals assume kinetic shapes as they grow at room temperature. As the temperature increases, the nanocrystals can assume equilibrium or kinetic shapes, depending on the temperature, their size, and the initial shape from which they evolved. The information in our study could be useful in efforts to devise processing routes to achieve selective nanocrystal shapes.

T. Yan, H. Zhang, and K. A. Fichthorn, *ACS Nano* **17**, 19288-19304 (2023).

J. Cui and K. A. Fichthorn (In progress).

55h Programming Soft Materials Assembly Via Inverse-Design

Chrisy Xiyu Du

Soft materials are crucial in many different forms of revolutionary technologies for their ability to self-assemble into intricate structures using a finite set of building blocks. Oftentimes, we can alter these building blocks with continuously tunable parameters, providing us with an almost infinite design space for potential materials properties. Because of this giant parameter space, forward approaches to sweep through parameters might not be the most efficient way to discover the building blocks for a specific material structure or property, and many inverse approaches have been developed in the past years. In this talk, I will discuss two different inverse-design strategies for building block selection for targeted assembly properties: digital alchemy, a method based on extended thermodynamics ensembles, and differentiable molecular dynamics utilizing JAX-MD. By utilizing these methods, we can inverse-design optimal particle shape for density-induced solid-solid phase transitions and best patchy particles for bulk and finite self-assembly.

55i My Excellent Adventures with Sharon: From Wtec (1999) to Mosdef (today)

Peter Cummings

I first met Sharon in September, 1999, so we are celebrating 25 years of friendship in this symposium. In this talk, I will reminisce, with references to science and engineering, our 25 years as professional colleagues and friends.

55j Developing Software and Stories in Service of Soft Matter Self-Assembly

Eric Jankowski

Accurately predicting the morphology of organic semiconductors or polymer nanocomposites depends on a combination of molecular models implemented using mathematical methods for specific computational hardware. Software engineering is therefore central to soft matter science, but there are no degree programs for soft matter software development specifically. Our field benefits from interdisciplinary perspectives, but a consequence of being at the interface between mathematics, physics, chemistry, engineering, and computer science is that trainees frequently perceive themselves as outsiders. We hypothesize that scientific software is more reliably and efficiently produced by teams of developers that feel professionally included. We deploy two related activities for improving inclusion in soft matter simulation: Personal storytelling and targeted trainings. We show that storytelling interventions help trainees both develop their communication skills as well as interrogate, and improve, their sense of professional identity and belonging. We show community-developed curriculum and teaching through the Institute for Computational Molecular Science Education both fills training gaps for soft matter software development and professional community-building. In concert, we demonstrate that a focus on engineering our trainees' mental models of themselves and their communities--the software that writes the software--improves the transparency, reproducibility, usability, and extensibility of molecular simulation software.

515b Probing Molecular Conformations of Lanthanide Binding Peptides at Air-Water Interface

Yiming Wang, Stephen Crane, Felipe Jimenez Angeles, Luis Ortuno Macias, Jason Marmorstein, E. James Petersson, Cesar de la Fuente-Nunez, Ravi Radhakrishnan, Kathleen J. Stebe

Lanthanides, or rare earth elements (REEs) are critical materials with unique luminescent, magnetic and catalytic properties that are exploited in a wide range of clean energy technologies which require REEs in high purity. However, REEs are difficult to separate from each other due to their similar size and electric valence. Thus, there is a growing need for green methods to separate these elements from mixtures. Lanthanide-binding tags (LBTs), engineered from the calcium-binding motif of bacterial proteins have been designed to bind selectively to Tb^{3+}. Furthermore, LBT adsorb at air-water interfaces, and thus can used as peptide surfactant for REE recovery by froth flotation methods. In order to understand the selective binding mechanism of LBT variants to different REEs (such as La^{3+}, Tb^{3+} and Lu^{3+}), here we have performed enhanced-sampling molecular dynamics simulations (Metadynamics) to investigate the binding free energy landscape of LBT peptides and various trivalent REE cations in aqueous solution and at air-water interface. For each LBT-REE system, we have identified 5 to 6 different metastable binding structures with small free energy differences ($\Delta\Delta G \leq 1$ kcal/mol), revealing the hidden complexity of the LBT-REE binding complex conformations. Importantly, we were able to quantify the influence of air-water interface on the molecular conformations of LBT-REE complexes, which is difficulty to probe from macroscopic experiments. This work paves road for determining quantitative structure-activity relationship of lanthanide binding peptides, as well as designing novel peptide surfactants for green and efficient REE extractions.

515c Interactions between AFM Tips and Biomembranes Via Multiscale Molecular Modeling

Saswati Panda, Zack Xu, Yuqi Huang, Gang-Yu Liu, Roland Faller

We study computationally the critical interactions between an Atomic Force Microscopy (AFM) tip and lipid membranes, a pivotal aspect in the advancement of nanoscale lipid constructs for applications in drug delivery systems and biosensors. We introduce a model that elucidates the complex molecular mechanisms underpinning the AFM tip-lipid interaction. The model comprises a multi-bilayer 1-palmitoyl-2-oleoyl-glycero-3-phosphocholine (POPC) lipid system on a substrate and a pyramidal-shaped amorphous silica Afm tip. Our study delves into the intricate forces at play at the lipid membrane surface during penetration, and the energy landscape that governs the mechanically induced rupture of lipid bilayers during AFM tip indentation. Furthermore, we extend our investigation to examine the local mechanics, the packing, and the size-dependent mechanics of the system. These phenomena are explored using coarse-grained MARTINI molecular dynamics simulation. The insights derived from this study are anticipated to significantly influence the future design of lipid-based nanoscale constructs.

515d Adsorption of miRNA on Silica Surfaces Probed with Metadynamics Simulations

Mahbuba Khanom, Vance Jaeger

Biomolecular adsorption onto silica surfaces is a ubiquitous phenomenon. Silica glass is not just common in the built environment but also in laboratory equipment and medical devices. Biomolecular adsorption to silica governs several scientifically interesting processes such as bio-preservation, chromatography, and bio-silification. However, the specific interactions between biomolecules and silica at the molecular level remain inadequately understood. Due to a limited number of experimental techniques available for probing the silica-biomolecule interface, there are considerable challenges in exploring the adsorption energy landscape and conformational preferences. However, understanding the arrangement of molecules at functional silica surfaces is crucial for designing new functional materials like chromatographic purification tags, preservation media, and non-fouling medical devices.

In this research, metadynamics (metaD) simulations were employed to investigate the adsorption behavior of a 21-nucleotide microRNA (miRNA21) onto silica surfaces under different interfacial pH conditions, solute salt concentrations, and salt species. The miRNA21 was selected because it is a cancer marker of clinical importance. Simulations yielded detailed molecular-level structures of the adsorption process. To enable exploration of various miRNA21 conformations on silica surface (adsorbed, desorbed, folded, and unfolded), a variation of metaD called parallel tempering metadynamics in the Well-Tempered ensemble (PT-metaD-WTE) was utilized. This technique allows for metaD to be applied alongside parallel tempering with a greatly reduced number of replicas. MetaD is a collective-variable (CV) based method in which a few slow conformational degrees of freedom are biased to enhance sampling. However, using more than two CVs becomes computationally expensive. Thus, PT-metaD-WTE uses thermal energy to overcome hidden energy barriers that the biased CVs do not consider.

Simulation results revealed a correlation between higher salt concentrations and increased binding strength, underscoring the crucial role of salt ions and water structuring in the adsorption phenomenon. Additionally, salt-mediated interactions between miRNA and silica were identified, indicating that the presence of a layer of cations in the electrical double layer (EDL) can mitigate repulsive energetic barriers between miRNA and the negatively charged silica surface. Furthermore, simulations demonstrated that acidic environments notably weaken the adsorption between miRNA21 and silica surfaces due to a reduction in salt-mediated interactions. Divalent cations were found to hinder adsorption compared to monovalent cations, likely due to their role in stabilizing the folded structure of miRNA21 in solution. These simulation results match previously published trends in other adsorption simulations. These findings underscore the significance of environmental conditions and salt-mediated interactions in determining the affinity of miRNA-silica adsorption. The mechanistic insights gained from this study offer valuable guidance for optimizing interactions between biomolecules and silica. Results have been compared to experimental observations of a recently developed silica-based biopreservation medium called CaRGOS and to other published experiments.

515e Molecular Dynamics Studies on the Effect of Silica Surface Characteristics on Ion Adsorption and Their Hydration Structures

Kaijie Zhang, Zhehui Jin

Ion adsorption on charged surfaces is a ubiquitous phenomenon of natural and engineered systems containing water and colloids, nanoparticles, or nanopores, which plays crucial roles in nanofiltration, colloidal assembly, and biological processes, *etc*. In subsurface porous media, silica is one of the most abundant rocks, which often carries negative charges due to the deprotonation of surface hydroxyl (-OH) groups. Ion adsorption on negatively-charged silica surfaces and their hydration structures at elevated salt concentrations are in the heart of various energy and environmental engineering applications, such as hydrocarbon recovery, CO_2 and H_2 storage, and hydrogeology.

The adsorbed ions on perfectly-smooth structureless charged surfaces are often conceptually divided into two regions, the so-called electric double layer: a *Stern layer* in which ions adsorb as inner- and outer-sphere surface complexes (ISSC, OSSC); a *diffuse layer* which screens the remaining uncompensated surface charge. Classical electrical double layer theories based on mean-field Poisson-Boltzmann equation are only valid for dilute solutions, arising from the negligence of ion size and ion-ion correlation effects. In addition, implicit solvent (water) models are often used. Therefore, near surface ion hydration structures cannot be revealed from these theories. By using molecular simulations, Bourg *et al.* (*Journal of colloid and interface science* **2011**, *360* (2), 701-715) confirmed that surface counterions close to smectite surfaces can form ISSC, OSSC, and Diffuse species with zero, one and two full hydration shells, respectively. Unlike layer-type clay minerals, deprotonation of silica leads to a highly localized and heterogeneous surface charge distributions. To the best of our knowledge, the effect of silica surface charge heterogeneity on ion adsorption and their hydration structures still remains unclear.

In this work, molecular simulations are conducted to study Na^+ and Cl^- adsorption and their hydration structures in the vicinity of silica surface at the salinity of 0.6 mol/L. The surface charge heterogeneity is represented by three adsorption sites, including adjacent sites (O_a^-), isolated sites (O_i^-), and hydroxyl groups (-OH). In line with Bourg *et al.*, ions can form ISSC, OSSC, and Diffuse species. However, we find that ion adsorption exhibits distinct structures around different adsorption sites. As ISSC1 and ISSC2 species, Na^+ ions exhibit preference for adsorbing around O_a^- than O_i^-, but Na^+ ions are more likely to adsorb around O_i^- as OSSC species. -OH groups are the less favored sites for Na^+ adsorption. On the contrary, the adsorption of Cl^- is mainly due to the tri-OH – Cl^- structure. Additionally, the ion pairs formed around O_a^- and O_i^- also contribute to Cl^- adsorption. Water molecules in the hydration shells of ions can form water-surface hydrogen bonds (HBs) and in-shell water-water HBs, which can facilitate the stabilization of adsorbed ions. The hydration structures of Na^+ and Cl^- ions near silica surfaces are highly dependent on water-surface HB interactions, resulting in several preferred hydration spots in the first/second hydration shells. The direct electrostatic interactions between ions and surface and water-surface HB interactions collectively influence ion mobility near surface. Our study from molecular perspectives emphasizes the importance of heterogeneity of silica surface charges and surface functional group characteristics on ion adsorption and their hydration structures. Our work can shed important insights into the fundamental understanding about ion-water-silica interactions which have paramount importances in gas storage and geohydrology.

515f Unriddling the Cationic and pH Effect in Alkaline Hydrogen Evolution Reaction on Pt

Zisheng Zhang, Aamir Shah, Yu Huang, Xiangfeng Duan, Anastassia Alexandrova

The alkaline water electrolysis is plagued by sluggish kinetics in cathodic hydrogen evolution reaction (HER) even on the platinum catalysts. The electrode/electrolyte interface in alkaline HER is also far more complex and controversial than the acidic case, due to presence of near-surface alkali metal (AM) cations, surface adsorbates, and different interfacial water structures that they lead to.

Herein, we conduct a density functional theory (DFT) study with explicit solvation, including static calculations, grand canonical DFT calculations, ab initio molecular dynamic (AIMD) simulation, and micro-solvation cluster calculations, to gain molecular level insights into the surface adsorption properties, solvation structure, and the dynamics at the Pt/water interface in presence of cations and surface OHad species, or at different pH values. Electrochemical impedance spectroscopy (EIS) and electrical transport spectroscopy (ETS) are also performed by experimental collaborators to probe the surface species and interfacial capacitance. Our integrated studies suggest that the AM cations play an indirect role in modifying HER kinetics: Smaller cations destabilize surface OHad to a lesser extent in the potential window of HER, leading to a higher OHad coverage on Pt surface. The highly polar surface OHad weakens the O-H in nearby water molecules, which can promote alkaline Volmer kinetics and hence boost the HER activity.

In addition, we combine cyclic voltammetry (CV) and electrical transport spectroscopy (ETS) approach to probe Pt surface at different pH and develop molecular level insights on the pH-dependent HER kinetics in alkaline media. The change in HER Tafel slope from ~110 mV/decade in pH 7–10 to ~53 mV/decade in pH 11–13 suggests considerably enhanced kinetics at higher pH. The ETS studies reveal a similar pH-dependent switch in the ETS conductance signal at around pH 10, suggesting a notable change of surface adsorbates. Fixed-potential calculations and chemical bonding analysis suggests this switch is attributed to a change in interfacial water orientation, shifting from primarily O-down below pH 10 to H-down configuration above pH 10. This reorientation weakens the O–H bond in the interfacial water molecules and modifies the reaction pathway, leading to considerably accelerated HER kinetics at higher pH. Our integrated studies provide unprecedented molecular level understanding of the non-trivial pH dependent HER kinetics in alkaline media.

The collection of works demonstrate the necessity and benefits of including explicit electrolyte and/or constant potential treatment in molecular dynamics simulations to properly understand the interfacial chemistry of even the simplest electrochemical reaction.

515g Computational Screening of Synergistic 2D Materials Impregnated with Deep Eutectic Solvents for Removal of Pharmaceuticals

Tarek Lemaoui, Ahmad Darwish, Fawzi Banat, Hassan Arafat, Inas M. AlNashef

Abstract

This study introduces a groundbreaking approach in the domain of water treatment technology, emphasizing the integration of a molecular modeling framework to pinpoint the most effective two-dimensional (2D) materials for the removal of pharmaceuticals from aqueous solutions. The research specifically focuses on graphene (Gr), graphene oxide (GO) and reduced graphene oxide (rGO), used with natural deep eutectic solvents (NADESs), a relatively underexplored class of eco-friendly solvents that offer potential for environmental remediation applications.

The initial phase of the research involved an extensive list of over 100 potential that can make NADESs. This foundational work set the stage for further experimental investigations. The next step involved employing Density Functional Theory (DFT) using the Materials Studio software with the General Gradient Approximation (GGA) and Dmol3 PBE functional. The results from these DFT calculations were used to generate COSMO files, which then served as inputs for the Conductor-like Screening Model for Real Solvents (COSMO-RS) analysis. The results of the DFT calculations were used to create COSMO files. These files then provided the basis for COSMO-RS analysis. Using COSMOthermX software, we calculated the activity coefficients at infinite dilution for the 2D materials including Gr, GO, and rGO, alongside water molecule, natural compounds and pharmaceutical pollutants, including antibiotics like Meropenem, Tetracycline, and Ciprofloxacin. The screening process concentrated on two pivotal performance metrics: affinity and selectivity at infinite dilution. These metrics are critical for determining the practical applicability of the materials in water treatment processes, where efficient and selective adsorption of contaminants is required.

Our findings demonstrate that GO has a significantly high affinity for a wide range of pharmaceutical substances, as evidenced by its activity coefficient at infinite dilution of -4,968. This performance markedly surpasses that of rGO and Gr, which recorded coefficients of 3,454 and 10,124, respectively, highlighting GO's enhanced capability in trapping these pollutants. Further investigation into NADESs revealed that among the terpene and terpenoid family compounds, thymol exhibited the highest affinity towards antibiotics with an activity coefficient of -6.487. This was significantly better compared to other tested natural compounds like menthol at 3.399, camphor at 3.035, and coumarin at 1.867. These findings underscore the potential of environmentally friendly compounds in the development of sustainable water treatment technologies.

In summary, this research provides profound molecular-level insights into removing pharmaceuticals from water using GO and NADESs. The outcomes not only contribute significantly to the understanding of 2D material-NADES interactions but also lay a robust foundation for future explorations. These future efforts will likely include the synthesis and characterization of synergistic 2D-NADES-based adsorbents tailored explicitly for the enhanced removal of pharmaceutical contaminants from aqueous media. The implications of such developments could be transformative for environmental protection strategies, offering scalable, practical solutions to one of the most pressing issues in water treatment.

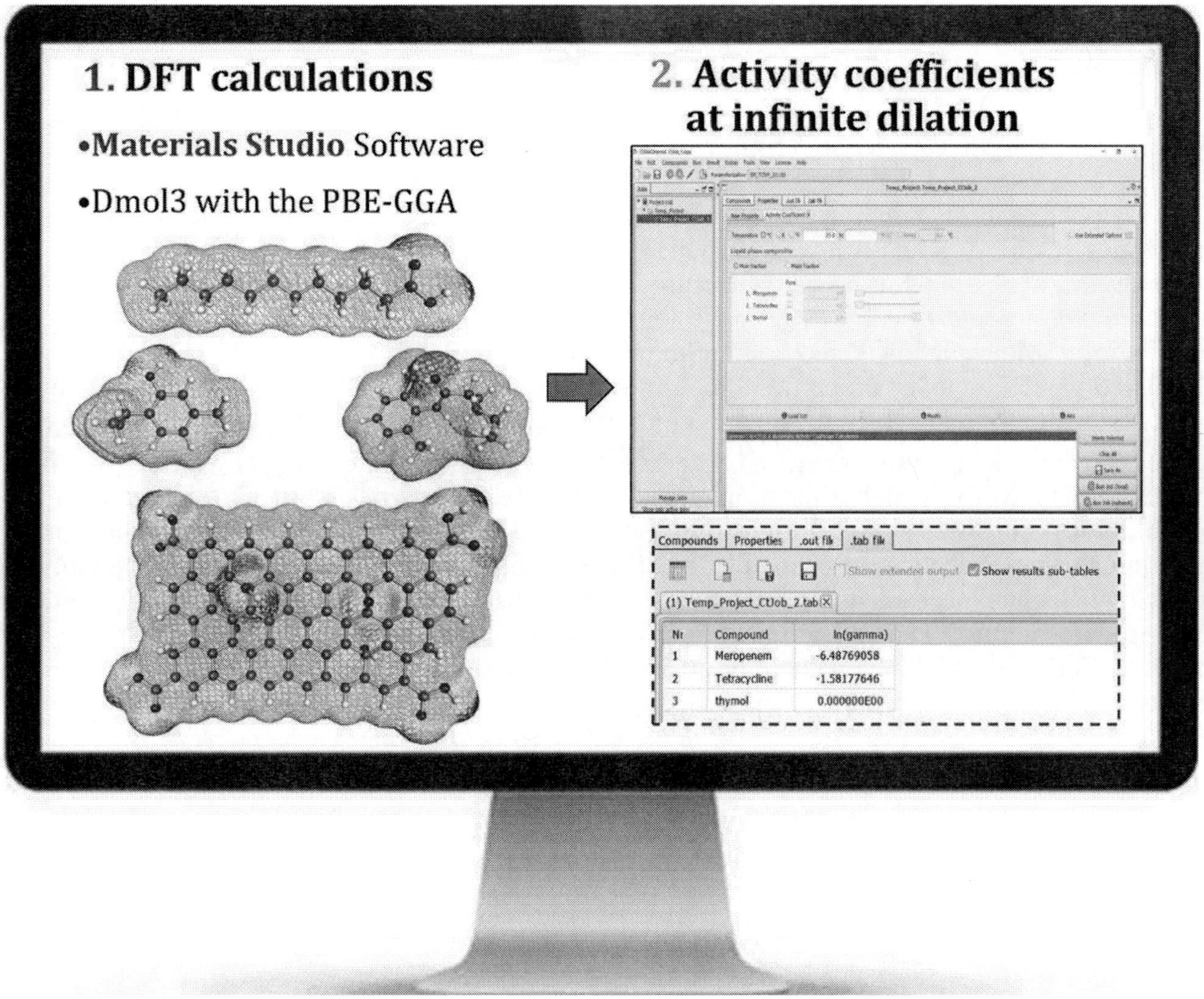
1. DFT calculations
•Materials Studio Software
•Dmol3 with the PBE-GGA
2. Activity coefficients at infinite dilation
Compounds
Properties
.out file
.tab file
Show extended output
Show results sub-tables
(1) Temp_Project_CtJob_2.tab
Nr
Compound
ln(gamma)
1
Meropenem
-6.48769058
2
Tetracycline
-1.58177646
3
thymol
0.000000E00

515h Aqueous Hydrogen in Confinement: Structural and Dynamical Properties Estimated from Simulations

Khang Bui, Tran LE, Gabriel D. Barbosa, Alberto Striolo

As the global focus on harnessing hydrogen energy to decarbonize diverse industrial sectors intensifies, the need for innovative approaches to store large volumes of this gas intermittently becomes more pressing. Large-scale underground hydrogen storage (UHS) might emerge as a viable solution. In this context, we investigate the solubility and diffusivity of hydrogen gas under confinement. Molecular Dynamics (MD) Simulation was employed for this project. The results are discussed in relation to the properties of confined water and compared to those obtained previously from our group for other gases in aqueous confinement, for instance, CO_2 [1], CH_4 [2], and H_2S [3]. In this research, nano-confinement and pore size variations of kaolinite material were proven to influence both hydrogen and water structural and dynamical properties. Particularly, hydrogen solubility was enhanced significantly with unexpected distribution of the gas and water molecules. In addition, the results also revealed interesting diffusivity behavior of aqueous hydrogen. Notably, the study unveiled the mechanism for fascinating observations of structural and dynamical properties of hydrogen gas within hydrated kaolinite pores. In conclusion, this research offers valuable insights and once verified experimentally, it could be attributed to optimizing strategies for large-scale UHS and enhancing the efficiency of hydrogen storage systems.

[1] A. Ali, A. Striolo, and D. R. Cole, 'CO2 Solubility in Aqueous Electrolyte Solutions Confined in Calcite Nanopores', J. Phys. Chem. C, ***2021****, 125(22), 12333-12341.*

[2] A. Phan, D. R. Cole, and A. Striolo, 'Aqueous Methane in Slit-Shaped Silica Nanopores: High Solubility and Traces of Hydrates', J. Phys. Chem. C, ***2014****, 118(9), 4860-4868.*

[3] S. B. Badmos, A. Striolo, and D. R. Cole, 'Aqueous Hydrogen Sulfide in Slit-Shaped Silica Nanopores: Confinement Effects on Solubility, Structural, and Dynamical Properties', J. Phys. Chem. C, ***2018****, 122(26), 14744-14755.*

515i Pressure-Driven Water Intrusion and Extrusion in Hydrophobic Nanopores

Abdelraheem Abdallah, Kolattukudy Santo, Alexander Neimark

Nanoporous particles with lyophobic properties hold great promise as materials for energy absorption and storage. When subjected to mechanical compression, these particles allow non-wetting solvents to intrude into nanopores, effectively converting and storing impact energy. Upon solvent extrusion, this stored energy can be released when the compression force is removed. While various nanoscale energy absorption systems (NEAS) have been explored experimentally, the molecular-level mechanisms governing the intrusion-extrusion cycles are not understood. To investigate this, we employ atomistic molecular dynamics (MD) simulations. Specifically, we investigate shock impact-induced intrusion and extrusion in hydrophobic cylindrical as well as slit pores using MD simulations. Our simulations, conducted for rigid and flexible nanopores of varying sizes under quasi-static and dynamic conditions, reveal three distinct stages of the compression process: (1) Initial elastic compression of the solvent, (2) Fluid intrusion into the pore upon reaching a threshold intrusion pressure, and (3) Elastic compression of the solvent upon pore-filling (Figure 1). By developing adequate computational models of flexible pores, we show that water intrusion and extrusion in flexible pores exhibit negative compressibility. We also observe that the extrusion of solvent from a completely filled nanopore, upon pressure release, is initiated by bubble nucleation. These simulation results are closely correlated with experimental observations.

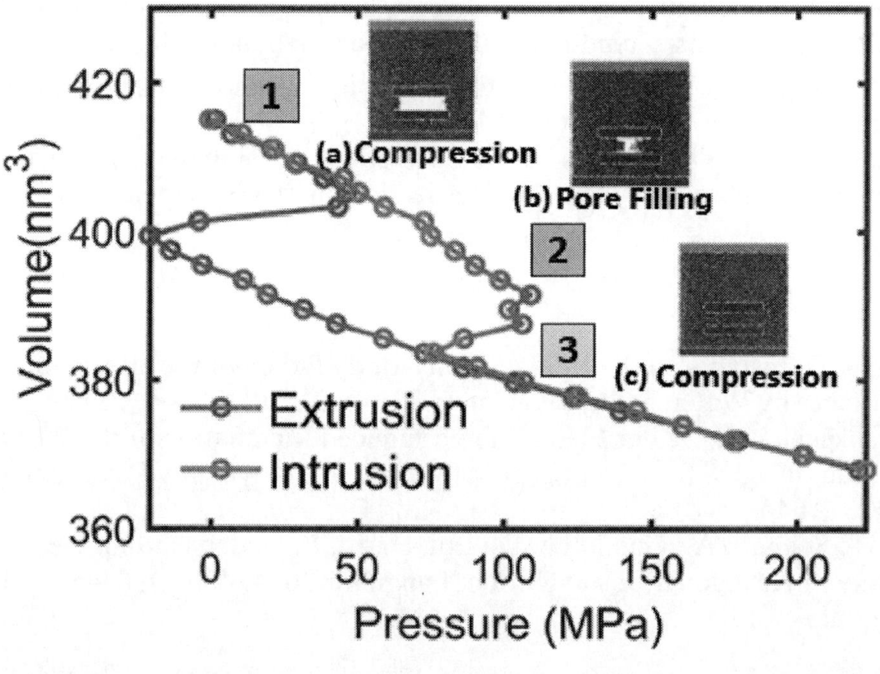

Figure 1. Change in volume of water, V, with pressure during the intrusion-extrusion cycle in hydrophobic flexible channel (a) Initial elastic compression of solvent (b) Pore filling at $P > P_{threshold}$ and (c) Elastic compression of solvent upon pore-filling

515j Using Polarizable Force Field-Based Molecular Dynamics Simulations to Model the Wetting of Hexagonal Boron Nitride Surfaces By Water

Shuang Luo, Rahul Prasanna Misra, Daniel Blankschtein

Hexagonal boron nitride (hBN) possesses outstanding attributes, including high thermal conductivity, wide band gap, and thermal stability, which have sparked significant scientific interest in this material. Many practical applications involving hBN, including seawater desalination, nanofluidic energy generation, biological sensing, and chemical separations, require that the hBN surface interacts with various liquids, with water being the most prevalent. Although polar water is expected to generate a finite electric field at an hBN/water interface, leading to the polarization of the charge distribution of the hBN surface, there remains a significant gap in understanding how this polarization controls the wetting ability of water on hBN surfaces.

With the above in mind, we developed a theoretical framework [1,2] that incorporates an all-atom polarizable force field to precisely capture the anisotropic polarizability tensor of hBN along with its interactions with water. This force field can self-consistently model the polarization of hBN induced by the electric field exerted by the water molecules using the classical Drude oscillator model. By carrying out Molecular Dynamics (MD) simulations, we investigated how the many-body polarization interactions influence the wetting properties of hBN surfaces by water. Our results show that this new force field predicts a water contact angle of 83.1° on hBN surfaces which is consistent with the recently reported experimental contact angle of 79.0° [3]. In contrast, using a Lennard-Jones (LJ) potential to model the water-hBN interactions, a commonly used approximation, predicted a significantly lower water contact angle of 67.8°[2]. We also found that incorporating the polarization response of hBN increases the number of water molecules adopting a 1 Leg orientation (with one of the hydrogen atoms pointing towards hBN) at the hBN surface, thereby increasing the interfacial entropy of water, an effect which is not observed when the polarization energy is modeled implicitly using the LJ potential.

References

1. Misra, R. P.; Blankschtein, D. Insights on the Role of Many-Body Polarization Effects in the Wetting of Graphitic Surfaces by Water. J. Phys. Chem. C 2017, 121(50), 28166–28179.
2. Luo, S.; Misra R. P.; Blankschtein D. Water Electric Field Induced Modulation of the Wetting of Hexagonal Boron Nitride: Insights from Multiscale Modeling of Many-Body Polarization. ACS Nano 2024, 18, 1629–1646.
3. Yang, F.; McQuain, A. D.; Kumari, A.; Gundurao, D.; Liu, H., Li, L. Understanding the Intrinsic Water Wettability of Hexagonal Boron Nitride. Langmuir 2024, 40(12), 6445–6452.

455a Accelerating *Ab Initio* Calculations of Chemical Bonding and Equilibria at Material Interfaces Using Machine Learning

Elizabeth Lee

The mechanism, rates, and chemical equilibria can be understood from free energy profiles of reactions at material interfaces, including heterogeneous catalysis, electrocatalytic reactions, and defect formation in functional materials. Predicting the energetics has involved a trade-off in accuracy between the enthalpic (i.e., interatomic potential) and the entropic (i.e., degrees of freedom) contributions to the free-energy. Traditionally, computational efforts in molecular reactions from *ab initio* calculations have been largely focused on the former. The latter, however, becomes increasingly important with increasing thermal effects and conformational entropy in many-body systems and high-temperature settings. Here, I introduce enhanced sampling approaches that combine machine-learning algorithms and electronic structure calculations, to bridge the gap between accuracy and system size for chemical reactions at material interfaces from first-principles. In particular, I present an enhanced sampling simulations tailored to *ab initio* calculations based on gaussian process models. I will demonstrate how the gaussian process method allows efficient sampling and direct computation of free energy landscapes with *ab initio* accuracy using examples from molecular reactions on transition metal catalysts and reactions under nanogap confinement, with applications in plastic waste conversion and chemical sensing.

455b Neural Network Wave Function Solver

Ni Zhan, William Wheeler, Kevin Huang, Peter Orbanz, Lucas Wagner, Elif Ertekin, Ryan P. Adams

By simulating atoms and electrons following the laws of quantum mechanics, we can obtain virtually all material properties of interest. Since the many-body wave function naively scales exponentially with number of particles, we use approximate methods to solve the time-independent Schrödinger equation. Variational quantum Monte Carlo is a polynomial scaling method that reaches high accuracy which depends on the form of the trial wave function. Neural networks have been used to parameterize the orbitals of the Slater determinant, and these ansatzes reached state-of-the-art accuracies on molecules. However, optimization and scaling become difficult as system size increases, and previous work has fixed all electrons' spins throughout training and focused on molecules over solids. We built the first neural network wave function for solids that has correct invariances to supercell translation and point symmetries of the space group. We show that symmetrizing over space group operations reaches ground-state energies with higher scalability than other ansatzes. We also allow for variable electron spin on the Bloch sphere, and include spin information in the neural network input, dot-product with orbitals, and spin sampling in the MCMC chain. Our continuous spin wave functions achieve better energies than previous NN wave functions on molecules, the periodic hydrogen chain, and periodic homogeneous electron gas. We show that spin sampling leads to largest improvements in physical systems that are expected to have frustrated spin. Our work opens the possibility of direct investigation of complex quantum phenomena in solids and spin systems.

455c Accelerated Discovery of High-Performance Organic Electrode Materials for Battery Applications Using an Interpretable Machine Learning Framework

Farshud Sorourifar, Madhav Muthyala, Jay Park, Shiyu Zhang, Joel Paulson

Conventional battery electrode materials rely on finite transition metals, which can diminish both the economic and environmental value of battery energy storage systems [1]. In contrast, organic electrode materials (OEMs), derived from abundant and readily available elements, offer a more sustainable energy storage solution. Despite the extensive structural diversity of OEMs, which enables tailored molecule design for specific properties, navigating this diversity can be challenging. The relationship between structure and battery properties is often unclear, and existing proxies for these properties usually necessitate costly ab initio calculations. Moreover, designing OEMs is complicated by the need for molecules to meet multiple objectives and possess feasible synthetic pathways [2].

Recently, there has been a shift from traditional first-principles models towards more versatile data-driven modeling paradigms. The latter methods are advantageous in that they can uncover complex (nonlinear) relationships between molecular encodings and material properties. A variety of different model structures, such as neural networks, random forests, and support vector machines, have been explored for molecular property modeling across different areas of chemistry and material science, e.g., [3, 4, 5]. Although these methods alleviate the need for domain expertise and/or trial-and-error Edisonian search methods, they typically result in limited insights into the underlying physics (due to their black-box nature) and often require large training datasets to generalize to unseen test data to an acceptable level.

To accelerate the discovery of high-performance OEMs (in terms of high energy density and long cycle life), we developed a new modeling framework referred to as SPARKLE (Symbolic Predictive Algorithm for Recognizing Key Molecular Elements). We applied SPARKLE to a large class (nearly 1 million) quinone derivatives by training interpretable predictive models for the specific energy and solvation energy in terms of a small set of learned molecular descriptors. These models were trained on a small set of density functional theory (DFT) data for ~100 quinones using a tailored symbolic regression method [6]. Each model is composed on four or less descriptors, all of which have an explainable structure that provides insights into what molecular features contribute most to the properties of interest. The learned models achieve high R^2 values on held-out test data as well as $R^2 >$ 0.7 on a set of >100,000 quinones whose properties were computed using a different DFT method [7]. Applying these models to a set of nearly 1M molecules (obtained using a molecular generator [8]), we were able to identify many candidate OEMs that had the potential to outperform the current state of the art. To more systematically filter this space, we developed a new synthetic accessibility metric that modifies the score from [9] to account for symmetry of the molecules. By analyzing the tradeoff between specific energy, solvation energy, and synthetic complexity, a final set of 27 candidates OEMs were synthesized and manufactured into zinc-ion cathodes for testing in a coin cell battery. We found that 62.9% of these candidates satisfied key battery performance targets, which is 3x higher than the collective success rate achieved by the Zhang lab over the past six years (only 20.8% success across 72 tested molecules chosen by expert intuition). Moreover, 5 of the 27 compounds were found to be stable up to 1500 cycles with a capacity retention of 70.4%, which is on par with the current state of the art OEMs. By accounting for synthesizability, SPARKLE-identified OEMs are also likely to have simpler synthesis pathways, which is critical for achieving low manufacturing costs in practice.

References:

[1] Liang, Y., Jing, Y., Gheytani, S., Lee, K. Y., Liu, P., Facchetti, A., & Yao, Y. (2017). Universal quinone electrodes for long cycle life aqueous rechargeable batteries. Nature materials, 16(8), 841-848.

[2] Liang, Y., & Yao, Y. (2023). Designing modern aqueous batteries. Nature Reviews Materials, 8(2), 109-122.

[3] Jin, W., Barzilay, R., & Jaakkola, T. (2018, July). Junction tree variational autoencoder for molecular graph generation. In International conference on machine learning (pp. 2323-2332). PMLR.

[4] Random Forest model with combined features: A practical approach to predict liquid-crystalline property

[5] Jorissen, R. N., & Gilson, M. K. (2005). Virtual screening of molecular databases using a support vector machine. Journal of chemical information and modeling, 45(3), 549-561.

[6] Ouyang, R., Curtarolo, S., Ahmetcik, E., Scheffler, M., & Ghiringhelli, L. M. (2018). SISSO: A compressed-sensing method for identifying the best low-dimensional descriptor in an immensity of offered candidates. Physical Review Materials, 2(8), 083802.

[7] Flores, S. P., Martin-Noble, G. C., Phillips, R. L., & Schrier, J. (2015). Bio-Inspired electroactive organic molecules for aqueous redox flow batteries. 1. Thiophenoquinones. J. Phys. Chem. C, 119(38), 21800-21809.

[8] Berenger, F., & Tsuda, K. (2021). Molecular generation by Fast Assembly of (Deep) SMILES fragments. Journal of Cheminformatics, 13, 1-10.

[9] Coley, C. W., Rogers, L., Green, W. H., & Jensen, K. F. (2018). SCScore: synthetic complexity learned from a reaction corpus. Journal of chemical information and modeling, 58(2), 252-261.

455d *Inverse Design of Materials with Globally Optimal Topology and Geometry through Mixed Integer Linear Programming (MILP)*

Daniel Laky, Alexander Smith

The rapid discovery and optimization of materials is key to the development of sustainable, safe, and economic advancements in technologies that can improve human health, create and store clean energy, and mitigate the impact of pollutants and contaminants. There have been great advances in the automated, high-throughput exploration of material design spaces that enable these developments but the massive scale of these design spaces means guidance is needed in their exploration. Inverse design allows us to quantify relationships between a material's intrinsic characteristics (e.g., geometry, topology, molecular structure) and its target properties (e.g., catalytic efficiency, energy storage capability). These identified relationships can then be exploited to design new materials that optimize a target property. These methods are particularly applicable to materials where shape is a key indicator of performance, such as metal organic frameworks (MOFs), zeolites, and self-assembled systems such as soft gels and polymer networks. In this talk we demonstrate how a materials topology and geometry, measured via Minkowski functionals, can be used for the inverse design and generation of new material structures through a mixed integer linear programming (MILP) approach.

Minkowski functionals, also known as intrinsic volumes, concisely quantify the topology and geometry of complex materials. For 2-Dimensional systems these characteristics represent the area, perimeter, and Euler characteristic of a material. These measures have been used to discover connections between a material's multi-scale structure and emergent physical properties such as porosity, viscosity, and permeability. The Minkowski functionals are concise, scalable in computation, and are directly interpretable. They also act as a powerful dimensionality reduction technique, allowing the use of simpler data centric models for prediction of material properties with equal or higher levels of accuracy (e.g., linear regression model as opposed to complex convolutional neural networks).

However, one drawback of the dimensionality reduction of the Minkowski functionals is the inability to reconstruct a material directly from these measures. Generative models, such as variational autoencoders and generative adversarial networks (GANs) can be used for this purpose, but they often require a tremendous amount of training data, do not directly account for physical constraints, and are black box models that are difficult to interpret. To address these challenges, we propose a framework leveraging mixed integer linear programming (MILP) for the generation of materials from a set of Minkowski measures.

Utilizing training data, we regress how a material's physical and chemical characteristics depend on the Minkowski measures. With material dependence as an objective function, our MILP framework is then able to create a material with globally optimal geometry and topology. This makes MILP an effective strategy for the rapid generation of material designs with specific Minkowski measures. It is also amenable to the direct incorporation of physically meaningful constraints to the final solution. This allows for the exploitation of the relationship between the Minkowski measures and physical properties of a material, allowing us to extrapolate and generate physically meaningful material structures with optimal properties. Furthermore, there is a potential for multiplicity of optimal solutions which can be useful in data augmentation tasks to aid in the training of other deep learning models. We demonstrate our MILP framework's utility on a set of real images and show effective reproduction of materials with complex geometry and topology. With this demonstration, we show the MILP is an effective tool in problems that can use Minkowski measures or functions of the

Minkowski measures to design topologically or geometrically optimal configurations.

[1] Fuhr, Addis S., and Bobby G. Sumpter. 'Deep generative models for materials discovery and machine learning-accelerated innovation.' *Frontiers in Materials* 9 (2022): 865270.

[2] Lyngby, Peder, and Kristian Sommer Thygesen. 'Data-driven discovery of 2D materials by deep generative models.' *npj Computational Materials* 8.1 (2022): 232.

[3] Noh, Juhwan, et al. 'Inverse design of solid-state materials via a continuous representation.' *Matter* 1.5 (2019): 1370-1384.

[4] Armstrong, Ryan T., et al. 'Porous media characterization using Minkowski functionals: Theories, applications and future directions.' *Transport in Porous Media* 130 (2019): 305-335.

[5] Slotte, Per Arne, Carl Fredrik Berg, and Hamid Hosseinzade Khanamiri. 'Predicting resistivity and permeability of porous media using Minkowski functionals.' *Transport in Porous Media* 131.2 (2020): 705-722.

[6] Mecke, Klaus R. 'Additivity, convexity, and beyond: applications of Minkowski functionals in statistical physics.' *Statistical Physics and Spatial Statistics: The art of analyzing and modeling spatial structures and pattern formation*. Berlin, Heidelberg: Springer Berlin Heidelberg, 2000. 111-184.

[7] Schröder-Turk, Gerd E., et al. 'Minkowski tensor shape analysis of cellular, granular and porous structures.' *Advanced Materials* 23.22-23 (2011): 2535-2553.

[8] Mecke, Klaus R., and Dietrich Stoyan, eds. *Morphology of condensed matter: physics and geometry of spatially complex systems*. Vol. 600. Springer, 2008.

[9] Smith, Alexander, and Victor M. Zavala. 'The Euler characteristic: A general topological descriptor for complex data.' *Computers & Chemical Engineering* 154 (2021): 107463.

[10] Jiang, Shengli, et al. 'Scalable extraction of information from spatiotemporal patterns of chemoresponsive liquid crystals using topological descriptors.' *The Journal of Physical Chemistry C* 127.32 (2023): 16081-16098.

[11] Smith, Alexander, et al. 'Topological analysis of molecular dynamics simulations using the euler characteristic.' *Journal of Chemical Theory and Computation* 19.5 (2023): 1553-1567.

[12] Laky, Daniel J., and Victor M. Zavala. 'A fast and scalable computational topology framework for the Euler characteristic.' *Digital Discovery* (2024).

455e Catberta: Catalyst Energy Prediction and Feature Analysis through Language Models

Janghoon Ock, Chakradhar Guntuboina, Amir Barati Farimani

Effective screening of catalysts requires models that can predict adsorption energy, a crucial indicator of reactivity. Current prevailing methodologies, graph neural networks (GNNs), require precise atomic coordinates to build graph representations and struggle to incorporate observable characteristics in a human-interpretable way. This study introduces CatBERTa, a Transformer-based energy prediction model leveraging the pre-trained RoBERTa natural language processing encoder. CatBERTa directly processes human-readable text, including compositions and structural descriptions of catalysts. This text can be formatted as either strings containing specific features or as natural language descriptions produced by generative language models such as ChatGPT. CatBERTa's predictions of adsorption energy, based on textual representations of initial structures, match the accuracy of conventional GNNs, such as SchNet and DimeNet++. Notably, in data subsets where GNNs achieve high prediction accuracy, CatBERTa consistently exhibits comparable accuracy, with a mean absolute error (MAE) of 0.35 eV. An ablation study demonstrates the importance of interacting atoms as descriptors for adsorption configurations, while deeming bond lengths and atomic properties less predictive. Moreover, the self-attention score from the Transformer architecture enables us to understand the model's focus on tokens related to adsorbates, bulk composition, and interacting atoms. Analysis of the self-attention scores and latent space demonstrates that the fine-tuning process shifts the model's focus towards text pertaining to the adsorbates and adsorption configurations. This work lays the groundwork for text-based prediction of catalyst properties, bypassing the need for graph representations and shedding light on complex feature-property relationships.

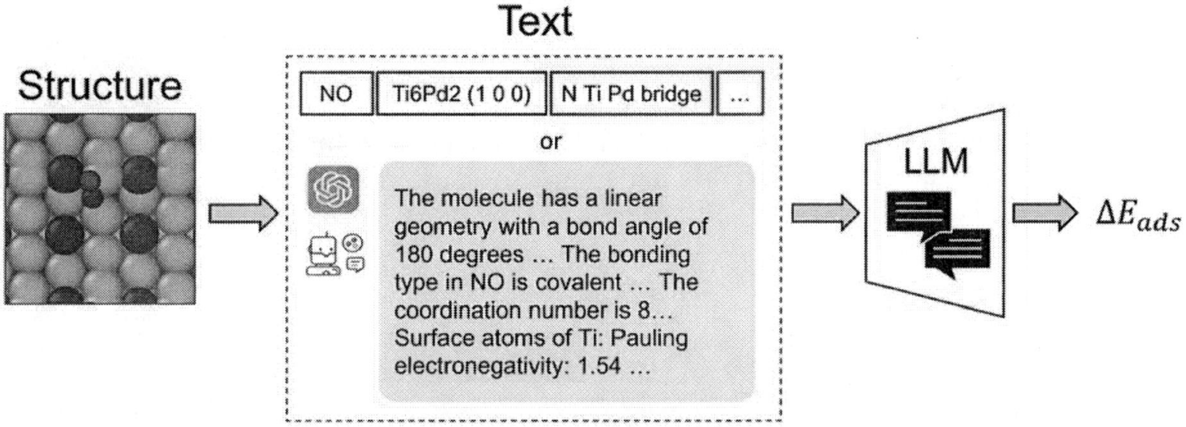

169bv Systematic Development of Machine Learning Interatomic Potentials for $MAPbI_3$ Perovskites

Nisarg Joshi, Nadia Intan, Marlo Zorman, Jim Pfaendtner

Obtaining potential energy surface (PES) of a molecular system has been shown to provide valuable insights of molecules, understanding their conformational changes and predicting their properties. Balancing the accuracy vs computational cost trade-off between first principle calculations and empirical force fields has been a major limitation to studying accurately large systems and complex phenomena with significant activation barriers. Machine learning potentials (MLP) have proved to bridge this gap and provide high accuracy from its first principle data at a fraction of the computational cost[1]. However, MLP simulations tend to encounter unexplored trajectories where its reliability and robustness is low. To overcome this challenge, the workaround surrounds gathering more training data with methods such as active learning[2]. In this study we propose a protocol for systematic development of MLP and apply them for $MAPbI_3$ perovskite systems. We demonstrate training protocols of MLP for aqueous systems and show their generalizability for interfacial and $MAPbI_3$ perovskites systems. We show using the development protocols how MLP can be used to understand the degradation mechanisms of perovskites in liquid water. Furthermore, we envision its applications for different surfaces to explore the extent to which they can be used for systems of varied size and different thermodynamic ranges.

References:

1. Unke OT, Chmiela S, Sauceda HE, et al. Machine Learning Force Fields. *Chem Rev*. Published online March 11, 2021. doi:10.1021/acs.chemrev.0c01111
2. Zhang, L., Lin, D.-Y., Wang, H., Car, R. & E, W. Active learning of uniformly accurate interatomic potentials for materials simulation. *Phys Rev Mater* **3**, 023804 (2019).

455g Machine-Learning Accelerated Simulation: Toward Design and Synthesis of Nanocarbon Materials

Rebecca Lindsey, Yanjun Lyu, Sorin Bastea

Carbon nanomaterials are of tremendous technological interest due to the manifold of properties they can exhibit. For example, nanodiamond is renowned its hardness and biological inertness and is found in applications spanning industrial lubricants to biological implants and drug delivery vehicles; graphitic nanoparticles including quantum dots, nanotubes, and graphitic nanoonions exhibit tunable electronic and optoelectronic properties that are being explored for quantum computing, energy harvesting, and electronic devices. However, exploration in this materials space remains nascent due to challenges associated with establishing efficient, scalable synthesis strategies, and navigating the massive design space, where both of these challenges stem from a limited understanding of phase behavior and phase transformations in carbon-rich materials. To overcome these challenges, we deploy quantum-accurate simulations using a new T/P transferable carbon model developed through our ChIMES physics-informed machine learning framework to enable quantum-accurate simulation on far greater spatiotemporal scales. Using this model, we provide predictions of the carbon melt line up to 100 GPa and explore kinetic effects during phase transformation. We also provide preliminary insights on how these findings translate to nano-scale materials. Ultimately, this simulation capability is helping to establish an accelerated pipeline for design, discovery, and synthesis of next-generation nanocarbon materials.

455h Using Large Language Model to Collect and Analyze Metal-Organic Framework Property Dataset

Jihan Kim

This research was focused on the efficient collection of experimental Metal-Organic Framework (MOF) data from scientific literature to address the challenges of accessing hard-to-find data and improving the quality of information available for machine learning studies in materials science. Utilizing a chain of advanced Large Language Models (LLMs), we developed a systematic approach to extract and organize MOF data into a structured format. Our methodology successfully compiled information from more than 40,000 research articles, creating a comprehensive and ready-to-use dataset. A machine learning model trained using both the experimental and the simulation data reveals the advantage of incorporating experimental data over relying solely on simulated data for enhancing the accuracy of machine learning predictions in the field of MOF research. As such, our study illustrates the importance of collecting massive amount of the available experimemntal data to guide the future directions of materials development.

455i Data-Driven Acceleration of Materials Development: Leveraging AI/ML to Scalably Bridge the Gap from Models to Materials

Mark Sullivan

Materials play a vital role across diverse industries, impacting areas from healthcare to consumer goods to automotive technologies and beyond. The process of researching, developing, and scaling up materials involves a multifaceted approach encompassing fundamental understanding of chemical and physical properties, chemical and/or materials science engineering, modeling ranging from molecular to macroscale, experimental workflows, and ultimately product development processes. Materials development requires a blend of fundamental-level insights and empirical knowledge around process development and optimization in order to analyze macroscopic properties and lead to the eventual development of a viable product that can be brought to market.

This presentation will explore methodologies for integrating information from molecular to macroscopic scales while incorporating both fundamental principles and applied domain expertise to address complex multiobjective challenges centered around materials development. Using an AI/ML approach, we will discuss how to leverage both fundamental and applied domain knowledge alongside data-driven insights to accelerate experimental processes with a focus on generating viable material products using small, information-rich datasets (<100 data points).

A case study from the materials industry will be used to illustrate the application of this approach in solving business-driven problems by iteratively using the relevant corpus of data to suggest statistically ranked candidate experimental procedures, experimentally testing those AI-generated protocols, reincorporating the resulting new data to improve models, and repeating this sequential learning process to arrive at novel materials that achieve complex performance targets. This work demonstrates the value of utilizing the entirety of relevant data on hand that an expert experimentalist will understand to have impact upon eventual material performance while not requiring explicit mathematical relationships between material properties or processing conditions when those relationships may not exist.

455j Simulation-Free, Two-Dimensional Histograms As Effective Adsorbent Representations for Machine-Learning Based Adsorption Predictions

Jair Fajardo-Rojas, Tsung Wei Liu, Tatiane Gercina de Vilas, Diego A. Gómez-Gualdrón

Gaining full control of adsorption phenomena in adsorbent materials could make a myriad of applications in numerous engineering areas possible, and metal-organic frameworks (MOFs) are among the most promising adsorbents to achieve such control. The promise of MOFs, however, is tied to their extreme chemical and structural tunability, which in turn gives rise to an overwhelmingly large number of possible MOFs. Thus, while molecular simulation has accelerated the study of adsorption in MOFs, even facilitating high throughput screening of hundreds of thousands of MOFs in some instances, it is likely that the exhaustive study of adsorption phenomena in general in MOFs requires "instantaneous" adsorption predictions such as those that machine learning (ML) models can provide.

It is known that the efficiency of ML models is highly dependent on the problem representation, where "meaningful" representations of the problem can aid the model learning, and even reduce the data burden for training. To this end, here we demonstrate the efficacy of two-dimensional histograms as adsorbent fingerprints that encode information about the MOF pore shape/size and interaction parameters associated with grids of adsorption sites in MOFs. Specifically, these histograms encode information about the MOF atoms closest to each site, and the (average) non-bonded interaction parameters associated with it. These histograms thus encode MOF properties that are independent of the adsorbed molecules but that yet are intrinsically related to adsorption behavior in general, facilitating their applicability along a variety of adsorption scenarios (instead of the usual customized MOF representation specific to a given adsorption problem). As a side benefit, these histograms can be generated for each MOF without the need of significant "domain knowledge," making them accessible to a wider research community.

To demonstrate their efficiency, we evaluate and discuss the efficacy of these histogram-based MOF representations to facilitate the application of data science techniques to three MOF screening scenarios requiring adsorption predictions: i) imbalanced learning to find adsorbent materials to work as NH3-extracting membranes in plasma reactors for green NH3 synthesis, ii) multitask learning to simultaneously predict the adsorption of multiple small molecules with the same model, iii) transfer learning scenarios to predict the adsorption of molecule "A" at condition "X", leveraging data for adsorption of molecule "B" at condition "Y".

For instance, we show that iterative, imbalanced learning using our proposed histogram representation found MOFs for NH3-selective membranes that were better that at least 70% of the structures that would have been assumed to constitute the top-200 MOFs if one would have relied on traditional hierarchical screening. As reference, note that in traditional hierarchical screening of adsorbents an early filter based on cheap calculations of adsorption Henry's constants is used to reduce the number of structures on which to perform full molecular simulations. Here, instead we use the structures coming out of that filter to "jumpstart" iterative imbalanced learning, which ultimately allowed us to find more promising MOFs while only performing full molecular simulation on less than 10% of the targeted MOF database. Additionally, we showed that the generated data was sufficient to yield meaningful data-driven design rules that could even be applied beyond MOFs to design adsorbents in general for these applications, .

455k Representation Learning for All-Silica Zeolites: Model Representations, Transfer Learning, and Multi-Task Learning

Yachan Liu, Ping Yang, Gustavo Perez, Aaron Sun, Wei Fan, Subhransu Maji, Peng Bai

Zeolites are widely used in selective chemical separation and catalytic processes. Their molecular-level shape selectivity is determined by the detailed 3D arrangement of framework atoms. We have recently developed and optimized ZeoNet, a representation learning framework that employs 3D convolutional neural networks (ConvNets) and a volumetric representation based on distance grids.[1] ZeoNet was trained to predict Henry's constants for n-octadecane adsorption in over 330,000 all-silica zeolites, achieving a correlation coefficient $r^2 = 0.97$ and a mean-squared error MSE of 3.8 in $\ln k_H$. In this study, we compared ZeoNet with other common representations, including 2D multi-view distance images, 3D multi-channel grids, point clouds based on atom coordinates and on surface sampling, and various graph networks. We found that 3D volumetric grids significantly outperformed other representations and employing a 3D ResNet18 architecture with single-channel distance grids achieved the best balance between computational efficiency and accuracy. Transfer learning was performed to predict adsorption for five other hydrocarbons with limited data, demonstrating that a pre-trained ZeoNet model could be effectively transferred to predict adsorption for these additional hydrocarbons with only hundreds of training samples. Finally, we also explored a multi-task learning problem, comparing different approaches to utilize available adsorption data for all six adsorption tasks simultaneously with a focus on obtaining the most transferable representation.

1. Y. Liu, G. Perez, Z. Cheng, A. Sun, S. Hoover, W. Fan, S. Maji, and P. Bai, "ZeoNet: 3D Convolutional Neural Networks for Predicting Adsorption in Nanoporous Zeolites," *J. Mater. Chem. A* **11**, 17570-17580 (2023).

455l High-Throughput Simulations and Machine Learning for Adsorption Processes

J. Ilja Siepmann, Yangzesheng Sun

The combination of high-throughput simulations and machine learning (HTS/ML) is a powerful tool that can aid in the discovery of nanoporous materials for gas storage and chemical separations. This talk will highlight applications of HTS/ML to (i) search for adsorbents with high temperature of maximum capacity for gas storage, (ii) optimize the desorptive drying conditions for water/alkanediol mixtures, and (iii) predict the spatial distribution of adsorbate molecules from three-dimensional energy grids of the adsorbent.

709a Computational Molecular Engineering for Advanced Materials

Yongjin Lee

In the dynamic domain of material science, the potential for fabricating advanced materials is vast, promising transformative advancements for sectors such as electronics, energy, and environmental science. These advanced materials, characterized by their significant diversity and adaptability, offer a virtually limitless array of structural variants by manipulating different molecular components. Exploring this extensive spectrum of potential structures necessitates employing cutting-edge computational methods to illuminate uncharted or overlooked regions of the chemical space and accelerate the discovery of materials with superior properties.

A particularly effective strategy to meet this need is the computational molecular engineering of materials. By virtue of advancements in molecular simulations and machine learning, this approach allows for the optimization and precise generation of structures targeted for specific applications. In our presentation, we will share our latest efforts in computational molecular engineering, specifically highlighting a machine learning strategy that combines Monte Carlo tree search with a recurrent neural network.

For practical demonstrations, we've applied our tool in virtual experiments to design 1) high-performance Metal-Organic Frameworks for methane storage and carbon capture, 2) Ionic Liquids with potential for flue gas (CO_2/N_2) and syngas (CO_2/H_2) separation, and 3) sustainable polyamides with enhanced properties. Our method has proven highly effective in the creation of promising materials.

Our approach can be easily adapted to other contexts by modifying the reward function to align with the desired performance property.

709b Generative Modeling of Polymer Melts for Solving Inverse Design Problems

Vinson Liao, Tristan Myers, Arthi Jayaraman

The ability to design novel polymeric materials with specified properties requires solving an, often, ill-posed problem via thorough exploration of a high-dimensional design space. Inverse design becomes especially challenging for polymeric systems as they exhibit for each design parameter, structural arrangements and order/disorder at various length scales that together lead to the desired material property. Choosing to use traditional molecular dynamics (MD) simulations alone to sample all of the relevant polymer chain configurations and various length scales of structural arrangements for every design parameter can be computationally expensive due to long relaxation times of chains. In this work, we introduce a framework to effectively explore the configurational space of polymer melts using machine learning and MD simulations. We utilize a variational autoencoder (VAE) trained on MD simulation trajectories of melts of coarse-grained polymer chains to learn chain configurations and embed them onto a continuous distribution of low-dimensional latent space variables. Our methodology can generate many chain configurations including ones (e.g., non-equilibrium extended configurations upon processing) that may be computationally difficult to sample via traditional MD simulations. Using a simple A-B coarse-grained bead-spring diblock copolymer and its design parameter space as an example, we demonstrate our approach. We will conclude our talk with a brief discussion of potential extensions of this framework specifically for design of next generation of novel, high-performance polymeric materials.

709c From Physics-Informed GPR for Carbon Material Supercapacitors to Data-Driven Material Discovery

Runtong Pan, Tao Wang, Sheng Dai, Mengyang Gu, Jianzhong Wu

Amorphous porous carbons are one of the most popular electrode materials for energy storage owing to their high electrical conductivity, large specific surface area and low-production cost. Doping carbon electrodes with heteroatoms such as nitrogen and oxygen proves effective to improve the performance of aqueous supercapacitors. However, the optimal condition of N/O doping remains elusive due to the complexity of the porous structure and electrochemical behavior. Both physics-based models and machine learning (ML) methods have been used to correlate the electrochemical behavior of carbon electrodes. While physics-based models face challenges in capturing the pseudocapacitance effects, empirical correlation of the capacitance with ML methods may lead to erroneous predictions because of the lack of physical inputs, especially at high charging-discharging rates for electrodes with high mesopore surface areas. In this work, we introduced a Gaussian process regression (GPR) method using a physical model as the prior knowledge to limit the coupling effects of different input parameters. The physics-informed GPR proves effective in characterizing the capacitive behavior of N/O-codoped carbon electrodes in both 6 M KOH and 1 M H_2SO_4 aqueous solutions. The quantitative descriptions of the electrode capacitance offer valuable insights for the synthesis of the carbon-based electrode materials with better performance. Guided by our machine learning results, we designed a general sodium amide activation strategy for the synthesis of hyperporous carbons from hypercrosslinked polymers. These hyperporous carbons exhibit a high specific capacitance of 610 F/g in an acid electrolyte, highlighting the power of ML methods for materials discovery.

709d A Computational Approach for Reliable Materials Discovery: Application to Green Refrigerant Discovery

Barnabas Agbodekhe, Edward Maginn

This talk describes a computational workflow for the reliable discovery of materials. We describe its application to the discovery of green refrigerants. Currently used refrigerant fluids are deleterious to the environment as they have high global warming potential, necessitating the need for new, green refrigerants. Conventional molecular discovery works involve screening existing databases of molecules like PubChem and assessing their utility experimentally or via semi-predictive engineering correlations. As an alternative, generative machine learning (ML) coupled with ML-based property predictors can be used to obtain novel molecules with desired properties. Mathematical programming in optimization-based frameworks for molecular discovery is also popular. All these approaches have strengths but also some notable weaknesses, which limit their use for de-novo materials discovery tasks.

The novelty of our work lies in developing an integrated approach that systematically draws upon the unique strengths of the methods mentioned above while concurrently mitigating weaknesses. Instead of screening existing public (but limited) databases or using generative ML models (which require enormous amounts of training data), we applied mathematical programming to exhaustively enumerate all feasible candidate molecules subject to well-informed design constraints. We used FineSMILES, a tool we developed for fast and exhaustive enumeration. FineSMILES collects packets of molecular fragments made up of pre-selected elements, screens the packets based on design constraints, and assembles them into SMILES strings. This collection of SMILES strings represents all or almost all feasible and desirable molecules subject to informed design constraints. This approach allows the easy enforcement of highly sophisticated design constraints, which allows for more control and precision in generating high-potential molecules. Applying the FineSMILES code to the refrigerant design problem generated hundreds of thousands of molecules, with more than fifty percent of them not present in the PubChem database. We then leveraged the unique strengths of ML for property prediction to screen generated molecules. We developed and applied ML-based tools for state-of-the-art and independent property predictions with uncertainty quantification for several thermodynamic, environmental, and safety properties of interest. We developed a property prediction approach that used predictions from simple group contribution (GC) models like the Joback GC model and molecular weight as input to Gaussian process (GP) models for accurate property prediction. We also used sigma profiles as a molecular descriptor and vector of input features to GP models for direct property predictions and, as a third approach, for predicting PC-SAFT model parameters for novel molecules. We used all three methods above independently to predict properties. We scored the reliability of predictions for each screened molecule based on the level of agreement between the independently predicted property values using the three methods above. We also considered the GP property prediction uncertainties as additional information for determining prediction reliability. Based on a consideration of thermophysical, environmental, and safety performance, several hundreds of molecules were identified as high-potential green refrigerants, with many of them not previously reported in the open literature. Retrosynthetic accessibility (RA) scores were used for preliminary assessment of the synthesizability of the high-performing molecules. A few tens of molecules were identified with favorable RA scores.

We then used high throughput molecular dynamics (MD) simulations using off-the-shelf interatomic potentials (or force fields) to further validate and assess the technical performance of the identified high-performing and 'likely synthesizable' molecules. MD simulations were used to ascertain the miscibility of discovered molecules for designing binary refrigerant mixtures. Relevant thermophysical properties of hundreds of feasible and advantageous binary mixtures across multiple compositions were predicted using high throughput MD simulations. Five of the most promising refrigerant mixtures, none of which have been previously reported and for which we recorded a high level of certainty in predicted performance, were proposed to experimental collaborators for testing.

This work details a novel integration and synergistic use of existing tools and methods to develop a robust computational molecular discovery workflow applied to the green refrigerant design problem. Hundreds of feasible, single-component candidates, not previously reported, were identified. Five novel green refrigerant mixtures were designed and recommended for experimental testing using some of the discovered molecules. The developed approach can be adapted to several other types of problems, especially those involving small molecule discovery and design.

709e Organic Electrolyte Discovery Using Reinforcement Learning for Aqueous Redox Flow Battery Energy Storage

Cameron Gruich, Jessica Tami, Siddhant Singh, Bichlien Nguyen, Jake Smith, Anne McNeil, David Kwabi, Bryan Goldsmith

Organic molecule design with multiple target properties is a combinatorically hard problem in that the vast structure-property outcomes of molecular scaffolds preclude exhaustive search from *ab initio* computational methods (e.g., density functional theory, DFT) and experimental characterization. Machine learning (ML) models have been used to generate molecules and explore structure-property relationships for applications in a high-throughput fashion, yet many molecular design problems tasked for generative ML have target property ranges that are rare events in the underlying chemical space [1]. In this study, we use reinforcement learning of generative ML models applied to one such rare-event sampling problem in molecular design—finding soluble, redox-active organic electrolytes for aqueous redox flow battery (RFB) energy storage [2].

Organic electrolytes have garnered interest for use in RFB energy storage applications, but these electrolytes must satisfy a set of key properties, such as a sufficiently negative reduction potential and stability under operating conditions. Quinone- and phenazine-based molecules have been explored for this application, but their stabilities are sub-optimal in terms of their capacity fade rate [3]. Based on our experimental measurements, we narrow our electrolyte search to a benzo[c]cinnoline (BZC) scaffold, a relatively underexplored class of molecule that we hypothesize as being a promisingly stable chemical space due to a ring-constraining bond across from the hydrazo bridge of the core scaffold. The stability of the core benzo[c]cinnoline scaffold is explored with DFT via thermodynamic measures of degradation for lab-tested molecules compared to their unconstrained azobenzene analogues.

A recurrent neural network is leveraged to first generate BZC molecules irrespective of their chemical properties (i.e., a prior model). Property-guided reinforcement learning is applied onto the model to train a new iteration of the model (i.e., an agent model) via the REINVENT 2.0 framework, which biases an ML model towards generation of preferable molecules via explicit penalties to the loss function [4]. For effective property reinforcement, we develop a structure-property ML model for redox potential trained on estimates of the potentials via thermodynamic cycle—the states of which we compute with DFT. Also, a structure-property ML model for solubility is trained by using solvation energy as a thermodynamic proxy. Screening results of the proposed candidates are discussed. This work ultimately leverages a generative ML workflow to efficiently explore the chemical space and propose promising benzo[c]cinnoline electrolytes for further downstream synthesis and characterization for aqueous RFB energy storage.

References:

1. Bilodeau, Camille, et al. 'Generative models for molecular discovery: Recent advances and challenges.' Wiley Interdisciplinary Reviews: Computational Molecular Science 12.5 (2022): e1608.
2. Luo, Jian, et al. 'Status and prospects of organic redox flow batteries toward sustainable energy storage.' ACS Energy Letters 4.9 (2019): 2220-2240.
3. Kwabi, David G., Yunlong Ji, and Michael J. Aziz. 'Electrolyte lifetime in aqueous organic redox flow batteries: a critical review.' Chemical Reviews 120.14 (2020): 6467-6489.
4. Blaschke, Thomas, et al. 'REINVENT 2.0: an AI tool for de novo drug design.' Journal of chemical information and modeling 60.12 (2020): 5918-5922.

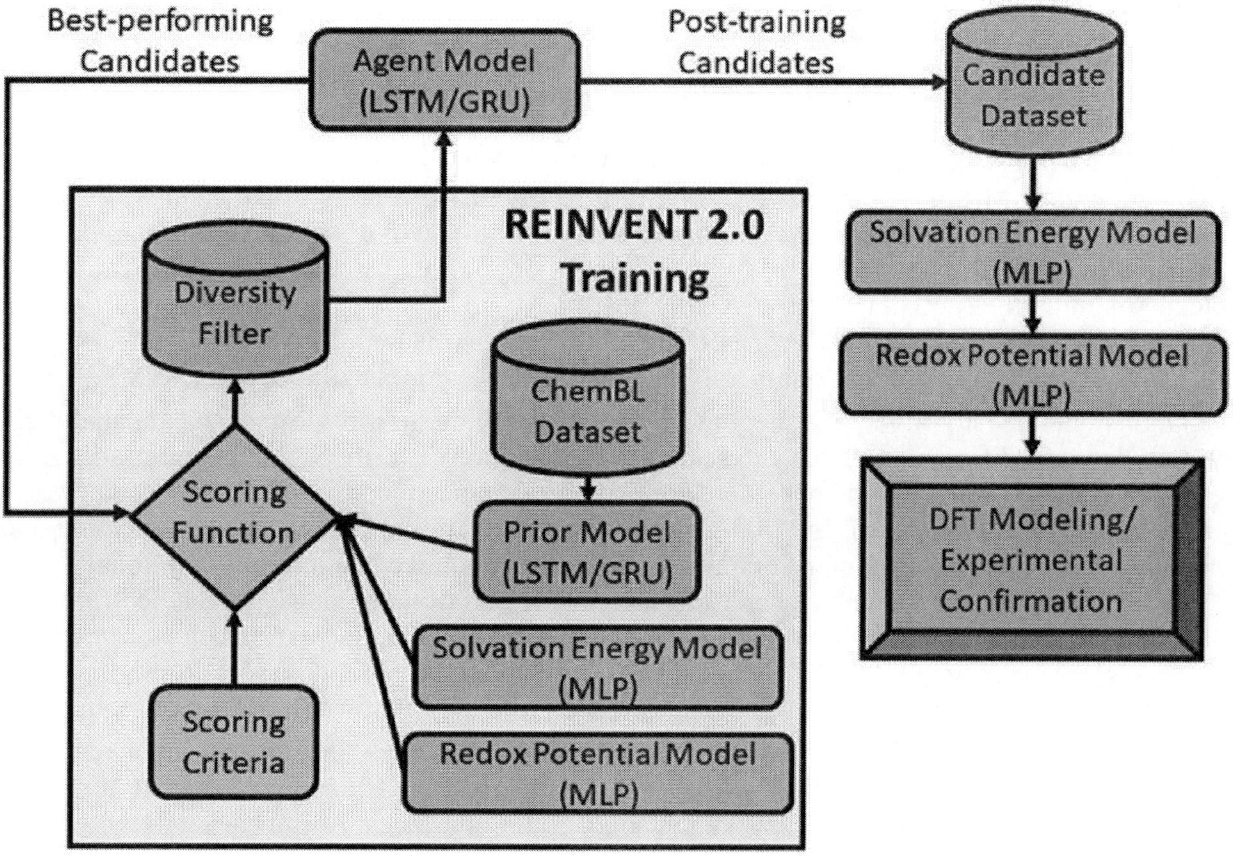
Best-performing
Candidates
Agent Model
(LSTM/GRU)
Post-training
Candidates
Candidate
Dataset
REINVENT 2.0
Training
Diversity
Filter
ChemBL
Dataset
Scoring
Function
Prior Model
(LSTM/GRU)
Solvation Energy Model
(MLP)
Scoring
Criteria
Redox Potential Model
(MLP)
Solvation Energy Model
(MLP)
Redox Potential Model
(MLP)
DFT Modeling/
Experimental
Confirmation

709f Materials Informatics for Structure-Property Relationships (MISPR) for Liquid Solutions

Rasha Atwi, Matthew Bliss, Nav Nidhi Rajput

In this talk, I will present a general purpose, community supported, open-source computational framework coined MISPR (Materials Informatics for Structure–Property Relationships) for guiding and accelerating materials discovery, optimization, and deployment of complex multicomponent liquid solutions. MISPR can tackle the twin challenges of automating and parallelizing multi-scale molecular simulations and developing machine learning frameworks for advanced scientific applications in chemistry, materials science and engineering. It is high-throughput multi-scale computational infrastructure that seamlessly integrates classical molecular dynamics (MD) simulations with density functional theory (DFT).[1] By enabling high-performance data analytics and coupling between different methods and scales, MISPR addresses critical challenges arising from the needs of automated workflow management and data provenance recording. The major features of MISPR include automated DFT and MD simulations, error handling, derivation of molecular and ensemble properties, and creation of output databases that organize results from individual calculations to enable reproducibility and transparency. I will describe newly developed fully automated DFT and MD workflows implemented in MISPR to compute various electronic properties such as nuclear magnetic resonance chemical shift, binding energy, bond dissociation energy, and redox potential with support for multiple methods such as electron transfer and proton-coupled electron transfer reactions.[2] The infrastructure also enables the characterization of large-scale ensemble properties by providing MD workflows that calculate a wide range of structural and dynamical properties in liquid solutions and at solid-liquid interfaces. At the backend, the infrastructure interfaces with the Gaussian [3] software which enables electronic structure calculations of chemical systems, and LAMMPS [4] open-source code for MD simulations. LAMMPS workflows allow the execution of MD simulations in different ensembles and analysis of the dumped trajectories for various dynamical and structural properties using another open-source codebase developed by our lab called **MDPropTools.**[5] **MDPropTools** is a powerful standalone in-house suite of Python-based post-processing routines, to perform statistical analysis of thermodynamic, structural, and dynamical properties of MD simulations of liquids and solid-liquid interfaces.[5] Using MISPR and MDPropTools, we published the first publicly available database, **ComBat** ~2,000 computed QC and MD properties for reported lithium sulfur battery electrolytes composed of solvents spanning 16 different chemical classes.[6, 7] Another open source database, **MICRO** consists of ~3000 computationally predicted properties for application in carbon dioxide electroreduction. Integrated machine learning models allow exploration of a larger chemical and parameter space at a significantly higher speed to establish structure-property relationships in multicomponent solutions. MISPR employs the methodologies of materials informatics to facilitate understanding and prediction of phenomenological structure-property relationships, which are crucial to designing novel optimal materials for numerous scientific applications and engineering technologies.

References

1. Atwi, R.; Bliss, M.; Makeev, M.; Rajput, N. N., MISPR: an open-source package for high-throughput multiscale molecular simulations. *Scientific Reports* **2022,** *12* (1), 15760.
2. Atwi, R.; Chen, Y.; Han, K. S.; Mueller, K. T.; Murugesan, V.; Rajput, N. N., An automated framework for high-throughput predictions of NMR chemical shifts within liquid solutions.

Nature Computational Science **2022,** *2* (2), 112-122.
3. Frisch, M. J.; Trucks, G. W.; Schlegel, H. B.; Scuseria, G. E.; Robb, M. A.; Cheeseman, J. R.; Scalmani, G.; Barone, V.; Petersson, G. A.; Nakatsuji, H.; Li, X.; Caricato, M.; Marenich, A. V.; Bloino, J.; Janesko, B. G.; Gomperts, R.; Mennucci, B.; Hratchian, H. P.; Ortiz, J. V.; Izmaylov, A. F.; Sonnenberg, J. L.; Williams; Ding, F.; Lipparini, F.; Egidi, F.; Goings, J.; Peng, B.; Petrone, A.; Henderson, T.; Ranasinghe, D.; Zakrzewski, V. G.; Gao, J.; Rega, N.; Zheng, G.; Liang, W.; Hada, M.; Ehara, M.; Toyota, K.; Fukuda, R.; Hasegawa, J.; Ishida, M.; Nakajima, T.; Honda, Y.; Kitao, O.; Nakai, H.; Vreven, T.; Throssell, K.; Montgomery Jr., J. A.; Peralta, J. E.; Ogliaro, F.; Bearpark, M. J.; Heyd, J. J.; Brothers, E. N.; Kudin, K. N.; Staroverov, V. N.; Keith, T. A.; Kobayashi, R.; Normand, J.; Raghavachari, K.; Rendell, A. P.; Burant, J. C.; Iyengar, S. S.; Tomasi, J.; Cossi, M.; Millam, J. M.; Klene, M.; Adamo, C.; Cammi, R.; Ochterski, J. W.; Martin, R. L.; Morokuma, K.; Farkas, O.; Foresman, J. B.; Fox, D. J. *Gaussian 16 Rev. C.01*, Wallingford, CT, 2016.
4. Plimpton, S. *Fast parallel algorithms for short-range molecular dynamics*; Sandia National Labs., Albuquerque, NM (United States): 1993.
5. https://github.com/molmd/mdproptools, **2022**.
6. https://github.com/rashatwi/combat.
7. Atwi, R. a. R., Nav Nidhi, Guiding Maps of Solvents for Lithium-Sulfur Batteries via a Computational Data-Driven Approach. . *Patterns http://dx.doi.org/10.2139/ssrn.4324048* **2023**.

169db Virtual High Throughput Screening of Car9 Isomers to Study Sequence-Binding Relationships

Aditya Sonpal, Jim Pfaendtner

The design of new solid-binding peptides (SBPs) is a non-trivial task driven by the relationship between the SBP sequence and the binding energy. This relationship is not deterministic from the amino acid (AA) sequence alone. This can be illustrated by Car9 (DSARGFKKPGKR). According to previous studies, with five positively charged AAs, Car9 binds strongly to a quartz surface (with –OH^- ions). However, not all of the five positively charged AAs participate in binding. The same has been found for a few selected mutants or sequence isomers of Car9. Therefore, we use Car9 and its ~20 million possible sequence isomers to study sequence-binding relationships. We narrow down this search space using unsupervised and semi-supervised clustering algorithms to end up with 1000 clusters with centers corresponding to 1000 peptides, which form a reasonable representation of the Car9 isomer space. We rank the 1000 resulting peptide sequences based on their sequence 'similarity' (calculated using metrics like Levenshtein and hamming distances) with Car9. We perform high throughput computational screening of the 1000 sequences using a new method of rapidly estimating binding energies developed by our group. We then determine if sequences 'similar' to Car9 have similar binding energies to the quartz surface. This would be a starting point for creating design rules for SBPs based on physics-agnostic similarity metrics. Understanding these relationships between the AA sequence and binding behavior is a step toward the inverse design of SBPs.

709h Data-Driven Discovery and Optimization of Small Molecules for Immune Modulation Via Machine Learning, Molecular Docking Simulation and Experiment

Oliver Tang, Aaron Esser-Kahn, Andrew Ferguson

Innate immune response plays a key role in protecting the body against pathogens and initiating the early stages of defense during infections. Therefore, precise modulation of the innate immune response is critical for maintaining effective pathogen protection while preventing excessive inflammation or immune-related diseases. This modulation is integral to the success of vaccines, which aim to stimulate protective immune responses with minimal adverse side-effects. Similarly, it is vital for the success of immunotherapies, as it can enhance immune stimulation and counteract the immune suppression within the tumor micro-environment. In our previous work, we have identified a short list of small molecules capable of modulating various innate immune responses by significant margins, up to more than 10-fold, utilizing a machine learning-driven pipeline to guide in vitro experimental screenings. Our next objective is to optimize the chemical structure of these small molecules to enhance their immune modulation capabilities and facilitate formulation into diverse delivery methods.

We have identified several potential targets within the innate immune system that our small molecules could interact with. Utilizing molecular docking simulations, we can simulate and elucidate the interactions between our small molecules and these targets. This enables us to guide the design of chemical modifications aimed at achieving enhanced binding affinity, ultimately leading to more potent immune modulation. By implementing a feedback loop between molecular docking simulations and machine learning, we can systematically explore a diverse range of chemical modifications to our initial hit molecules in an automated and efficient manner. This iterative approach yields modified chemical structures with strong binding affinities to potential immune targets, which can then be subjected to experimental validation. Importantly, our machine learning model incorporates considerations of formulation capabilities, such as synthetic accessibility, as well as water and oil solubility. This allows the model to prioritize modifications that are likely to exhibit favorable deliverability characteristics. By biasing towards molecules with optimal formulation properties, we increase the likelihood of successful experimental validation and subsequent clinical application.

This integrated simulation and machine learning optimization strategy not only accelerates the discovery of promising small molecule immune modulators but also enhances our ability to tailor these molecules for practical use in therapeutic interventions. The synergy between computational simulations, machine learning, and experimental validation represents a powerful paradigm for advancing the development of novel vaccine adjuvants and immunotherapeutic strategies, ultimately aiming to improve immune responses against pathogens and tumors while ensuring safety and efficacy in clinical applications.

709i Topological Data Analysis and Molecular Simulations for Identifying Nanostructure Transitions in Ionic Liquids

Lisa Je, Reid Van Lehn, Victor M. Zavala

Ionic liquids (ILs) are room temperature salts that have garnered attention as versatile electrolytes and designer solvents owing to their highly modifiable properties based on their cation and anion constituents. Unlike conventional solvents, ILs can self-assemble to exhibit nanostructures that can significantly impact IL properties such as polarity, hydrophobicity, and heterogeneity characteristics [1, 2]. These changes occur due to alterations in ion arrangements, inducing different molecular domains [3]. Variations in ion structuring can lead to incompatible molecular arrangements and subsequent segregation into different phases on a small scale [4], [5], [6]. However, the molecular-scale structure of neat ILs remains inadequately characterized and difficult to predict, hindering their rational application. Molecular simulations offer a viable approach to model the dynamic formation of IL structure but quantifying structural properties remains challenging the dynamic nature of IL structure, particularly in relation to its underlying molecular architecture.

In this work, we demonstrate a novel post-processing method that uses topological data analysis to convert molecular simulation data into a density field resulting in a Euler characteristic signature curve that identifies an IL structural transition. Atomistic molecular dynamics (MD) simulations were employed to quantify structure-dependent variations in 1-alkyl-n-imidazolium tetrafluoroborate (n-Mim BF4) pure ionic liquids as the dataset. For this specific imidazolium cation series, increasing the alkyl chain length increases the solvophobic interactions between the nonpolar alkyl chains and overpowers the electrostatic interactions of the cation and anion which in turn create a different nanostructure of ions. Literature has shown evidence that in the imidazolium cation family, the length of the alkyl chain at C=4 regime indicates significant change in nanostructure is observed computationally [7], [8], [9]. Nanostructure separation within the imidazolium cationic family is crucial, as it allows for precise tuning of ionic liquid properties based on variations in ionic structures. For each datapoint, we converted the MD trajectory into a density field resulting in a Euler characteristic curve where each n-Mim BF4 (n = 2, 3, 4, 5, 6, 8, 10, and 12) obtained their own "signature".

We demonstrate that the Euler characteristic curve identifies revealed a nanostructure transition in the IL imidazolium homologous series at n = 4 with less computational resources. At low alkyl chain lengths, we observe the ions dispersed as a homogenous nanostructure, but as the length of the alkyl side chain length increases, we observe a structural evolution where the nanostructure clusters heterogeneously with polar and non-polar regimes agglomerated. Investigating these effects of varying alkyl chain lengths on specific nanostructure transitions will contribute valuable insights to the literature on imidazolium cationic families and other IL systems.

[1] O. Nordness and J. F. Brennecke, 'Ion Dissociation in Ionic Liquids and Ionic Liquid Solutions,' *Chemical Reviews,* vol. 120, no. 23, pp. 12873-12902, 2020.

[2] S. Chen *et al.*, 'Ionic liquid clusters: structure, formation mechanism, and effect on the behavior of ionic liquids,' *Physical Chemistry Chemical Physics,* vol. 16, no. 13, pp. 5893-5906, 2014.

[3] K. Shimizu, C. E. S. Bernardes, and J. N. Canongia Lopes, 'Structure and Aggregation in the 1-Alkyl-3-Methylimidazolium Bis(trifluoromethylsulfonyl)imide Ionic Liquid Homologous Series,' *The Journal of Physical Chemistry B,* vol. 118, no. 2, pp. 567-576, 2014.

[4] Y. Zhang *et al.*, 'Effect of ion structure on the nanostructure and electrochemistry of surface active ionic liquids,' vol. 630, pp. 931-939, 2023.

[5] D. W. Bruce *et al.*, 'Nanosegregation and Structuring in the Bulk and at the Surface of Ionic-Liquid Mixtures,' *The Journal of Physical Chemistry B,* vol. 121, no. 24, pp. 6002-6020, 2017.

[6] M. A. A. Rocha *et al.*, 'Alkylimidazolium Based Ionic Liquids: Impact of Cation Symmetry on Their Nanoscale Structural Organization,' *The Journal of Physical Chemistry B,* vol. 117, no. 37, pp. 10889-10897, 2013.

[7] S. M. Urahata and M. C. C. Ribeiro, 'Structure of ionic liquids of 1-alkyl-3-methylimidazolium cations: A systematic computer simulation study,' *The Journal of Chemical Physics,* vol. 120, no. 4, pp. 1855-1863, 2004, doi: 10.1063/1.1635356.

[8] J. N. A. Canongia Lopes and A. A. H. Pádua, 'Nanostructural Organization in Ionic Liquids,' *The Journal of Physical Chemistry B,* vol. 110, no. 7, pp. 3330-3335, 2006.

[9] L. I. Vazquez-Salazar, M. Selle, A. H. de Vries, S. J. Marrink, and P. C. T. Souza, 'Martini coarse-grained models of imidazolium-based ionic liquids: from nanostructural organization to liquid–liquid extraction,' *Green Chemistry,* vol. 22, no. 21, pp. 7376-7386, 2020.

310a Denoising Diffusion Models Meet Statistical Mechanics for Protein Protein Docking

Sudeep Sarma, Lee-Shin Chu, Jacky Chen, Jeffrey J. Gray

Understanding how proteins interact is crucial in most biological processes as protein-protein interactions are involved in nearly all cellular functions in living organisms, from enzyme catalysis to signaling and gene regulation. Emerging protein-protein docking methods using generative AI provide a fast way to predict protein complex structure and its binding affinity given the structures of its unbound monomeric partners.

Traditional protein docking approaches comprise of two stages: (1) a sampling step to generate ensembles of candidate docked structures and (2) a ranking step to evaluate the docked poses from the sampling step. Recent breakthroughs in deep learning, particularly in likelihood-based models such as denoising diffusion probabilistic models, have been proposed for testing the rigid-body protein-protein docking task. The main idea behind a diffusion process is transforming a data distribution to a Gaussian prior over a series of time steps and learning a *score function*, which is the gradient of the log probability density function $\nabla_x \log P_t(x)$. We can then use the learned score function to sample from the underlying probability distribution of the data. Using this approach, we have developed a rigid body docking method: DockDiffusion. DockDiffusion was trained by adding rotation and translation noise to docked protein complexes, which the model then learns to reverse. During inference, the model takes two unbound monomers as input and solves a reverse diffusion process using a stochastic ordinary differential equation.

A grand question in deep learning-based protein structure prediction methods is whether these models have learned any thermodynamics associated with protein folding. To approach this question, we have estimated the change in exact likelihood ($\log P_0(x) - \log P_T(x) = -\Delta\log P$) of a pre-trained diffusion model for the protein-protein docking task and correlated it with the Rosetta energy function. We draw inspiration from statistical mechanics, where the thermodynamic potential (Gibbs energy in the case of an isobaric-isothermal ensemble) is proportional to the log of the probability distribution of observing a system in a particular state. Hence, $-\Delta\log P$ in the diffusion-based protein docking task might be a measure of the binding energy between two protein partners in contact. This exact likelihood estimation from a diffusion-based protein docking model will serve us two purposes: (1) first, it will help us to rank candidate poses without an external energy function, and (2) second, it will help us interpret whether diffusion-based docking methods have learned generalizable physical principles. We observed a robust correlation between $-\Delta\log P$ and the Rosetta energy function in one test case, whereas in another test case, no discernible correlation was detected between these two quantities. These results suggest that likelihood estimation does not correlate with Rosetta Energies and such models might have not yet "learned" generalizable physics. Answering fundamental questions like why and how deep-learning models work for protein design and modeling will help us build more robust tools for biomolecular engineering.

310b Development of *Ab-Initio*-Based Machine Learning Models to Study Chemical Reactions in Molten Alkali Carbonates and Hydroxides

Dina Kussainova, Athanassios Panagiotopoulos

Alkali-metal carbonates and hydroxides are widely used in energy and environmental applications due to their appealing properties. At high temperatures, pure carbonates and hydroxides decompose into vapor carbon dioxide and water, respectively. Moreover, experiments show that hydroxides can capture carbon dioxide by producing carbonates and water. Modeling these chemical reactions in the pure and mixture systems of molten carbonates and hydroxides can be crucial in better understanding their reaction mechanisms and designing their applications. Chemical reactions can be accurately predicted using *ab initio* molecular dynamics (AIMD) simulations, but high computational costs demanding small system sizes and time scales make this approach not feasible for our study. Classical molecular dynamics (MD) simulations, on the other hand, can efficiently predict system properties at different conditions using large system sizes and long time scales. However, they are not able to model chemical reactions in the systems except via detailed, predefined reaction mechanisms making the observation of new reactions and unforeseen products challenging. Recent advances in machine learning technologies can help overcome these challenges by training deep neural networks on quantum chemical data. The machine-learning models have shown good results in retaining the accuracy of the underlying ab initio methods while being comparable in efficiency to classical MD simulations. In the current work, we generate machine-learning models based on the AIMD simulations for pure lithium carbonates and hydroxides as well as their mixtures to study chemical reactions in liquid phase and vapor-liquid equilibria. We perform direct coexistence simulations to analyze the reactive multiphase behavior of systems at different conditions. We evaluate our results in terms of system composition, equilibrium constants, and partial pressures of vapor components. Additionally, we study system and volume size effects on the obtained results and analyze how the addition of new components affects the equilibrium constants. In general, our machine learning model predictions are in good agreement with available experimental results.

310c Scaling Molecular Modeling to Millions of Reactions with Neural Network Potentials

Dylan Anstine, Olexandr Isayev

Developments in high-throughput experimentation, automated chemistry platforms, and chemical generative models have created an urgent need to rapidly predict reaction outcomes so that synthetic planning and evaluations of degradability can match the emerging pace of molecule/material discovery. Machine learned interatomic potentials (MLIPs), whereby potential energy surface representations are learned from datasets of first-principles calculations, present an attractive opportunity to overcome the expensive and/or time-consuming tasks required to characterize and refine reactions experimentally or with quantum chemistry models. In this talk, I will introduce two models, RxnAIMNet and AIMNet2-Pd, that illustrate how MLIPs can be used to predict thermodynamics of general organic reactions and carbon-carbon cross couplings that are catalyzed by Pd organometallic complexes. By learning from a newly constructed and exhaustive dataset of ~10 million molecular conformers, RxnAIMNet is shown to reliably perform minimum energy pathway searches, transition state optimization, and intrinsic reaction coordinate calculations, leading to predicted activation barriers within ~2 kcal/mol of reference range-separated hybrid density functional theory (DFT) calculations. Despite our reactive training dataset being constructed from systematic enumeration of bond breaking and forming rules, we observe that RxnAIMNet implicitly learns key reaction mechanisms—such as Diels-Alder, hydrogenation, and triazole formation —supporting its broad transferability across general organic chemistry. To meet the needs of high-throughput reaction characterization, we introduce a method for batched nudged elastic band calculations, which is shown to allow RxnAIMNet to identify ~350,000 minimum energy pathways daily on a single medium-end GPU. Beyond concerted organic chemistry pathways, I will discuss the recently introduced AIMNet2-Pd, a model for 3D molecular modeling of Pd-containing organometallics, representing crucial first steps toward the underexplored areas of MLIPs for hybrid organic-inorganic species and systems undergoing chemical reactions. Considering Suzuki coupling as a representative reaction class, we demonstrate that the intermediate structures and main mechanisms (oxidative addition, transmetalation, and reductive elimination) can be evaluated for diverse ligands at a rate ~10^6 times faster than the standard quantum chemistry calculations used in homogenous catalyst screening. Taken altogether, these two models support an enhanced ability, especially on accelerated computing hardware, to tailor synthetic outcomes and design new reaction pathways by traversing approximate machine learned potential energy surfaces.

310d Two-Point Correlation Function Based Statistical Approach for Tracking Morphological Changes in Nanoparticle Systems

Puneet Koli, Rajdip Bandyopadhyaya, Abhijit Chatterjee

During the synthesis of nanoparticles and nanoclusters, their size, shape, morphology, porosity and number density can vary as a function of time. The exact dynamics depends on the synthesis conditions, and questions regarding the sensitivity towards the synthesis parameters arise. Experimental imaging techniques and stochastic modeling methods can generate spatially-resolved snapshots of the nanoparticle system. Given the randomness inherent in these images, directly comparing two snapshots may not convey the relative importance of the underlying factors. In this context, statistical techniques are better suited for tracking the dynamical transformation in such systems. In this talk, I introduce a technique based on the 2-point correlation (2PC) function for studying morphological transformations in nanoparticle systems. For our model system, the nanoparticle agglomeration dynamics is simulated using the kinetic Monte Carlo (KMC) method. The progress of the nanoparticle is quantified in terms of percentage transformation. The effect of the nanoparticle synthesis conditions on the morphology can be assessed. Additionally, rather than use multiple nanoparticle features (size, shape, number density, etc.) in tandem, as is conventionally done, the 2PC function can single-handedly characterize the nanoparticle evolution. Furthermore, quantities such as the nanoparticle size, shape and spacing are shown to be related to the 2PC function. This validates that our 2PC method as superior approach compared to the existing particle size measuring techniques, for characterizing nanoparticle structures.

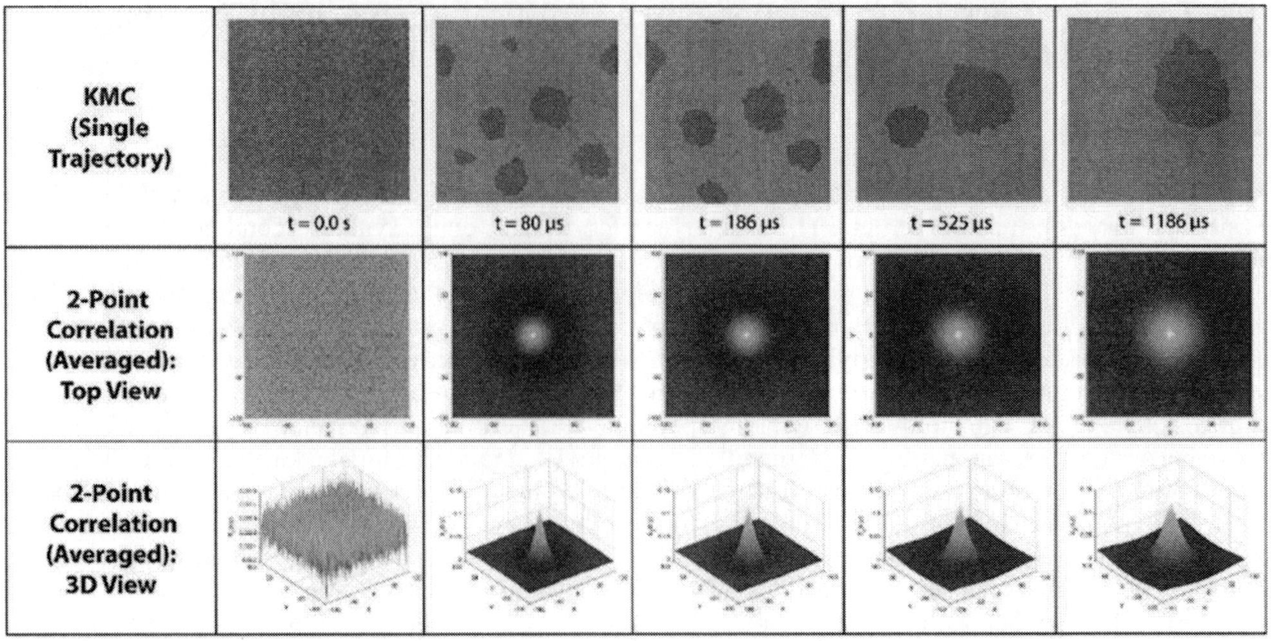
KMC (Single Trajectory)
t = 0.0 s
t = 80 μs
t = 186 μs
t = 525 μs
t = 1186 μs
2-Point Correlation (Averaged): Top View
2-Point Correlation (Averaged): 3D View

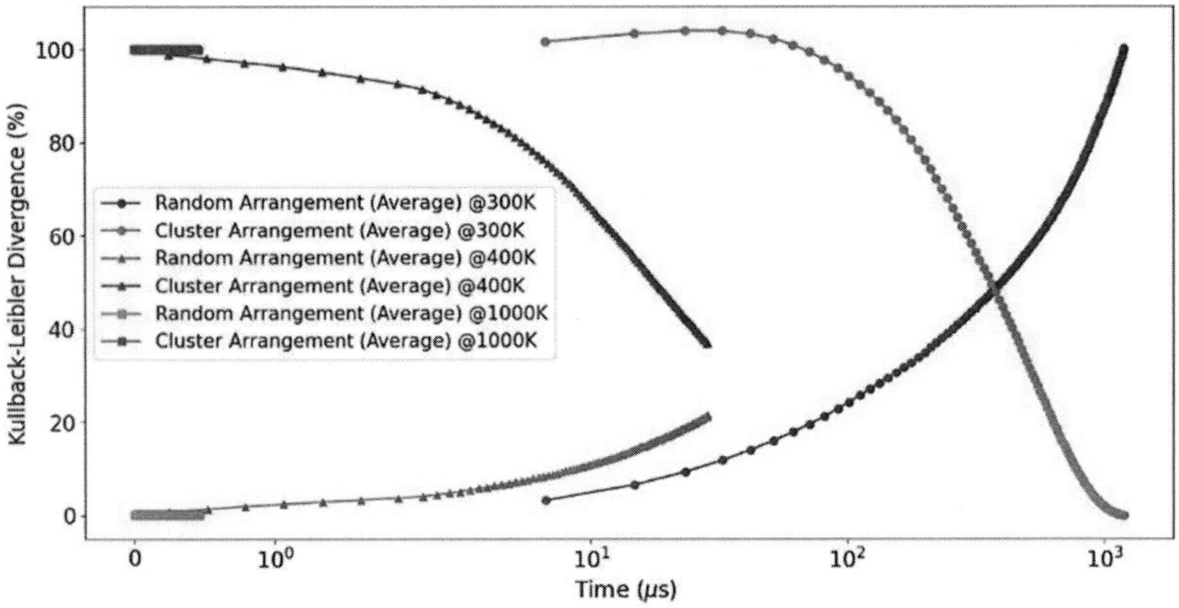
Kullback-Leibler Divergence (%)
Time (μs)
Random Arrangement (Average) @300K
Cluster Arrangement (Average) @300K
Random Arrangement (Average) @400K
Cluster Arrangement (Average) @400K
Random Arrangement (Average) @1000K
Cluster Arrangement (Average) @1000K

310e Efficient Generation of Barrier Crossing Trajectories Using Approximate Brownian Bridges

Vivek Narsimhan, George Curtis, Doraiswami Ramkrishna

We examine conditioned continuous random walks which reach one region before another. These conditioned processes are used to efficiently sample rare trajectories for barrier crossing events. The processes are conditioned using a Brownian Bridge construction, resulting in near perfect sampling efficiency without accruing any loss in the conditional statistics of the process. The construction requires the hitting probability or committer function, which is a solution to the Backward Fokker-Planck (BFP) equation, a PDE that can be difficult to solve through general means. Therefore, we derive a 1D analytical approximation using the asymptotic properties of the BFP which approaches the true solution as barrier height increases. Trajectories generated via this approximate bridge are then shown to result in accurate conditional statistics when used in conjunction with importance sampling, even in the case when potential energy barriers are not large. We demonstrate this methodology by simulating rare events in a stochastic chemical reaction network (Schogl reaction) with multiple steady states. This methodology shows great promise for future implementation for simulating rare barrier crossing events for a wide variety of physical processes.

310g Monte Carlo Simulations of Water Pollutant Adsorption at Parts-per-Billion Concentration: A Study on 1,4-Dioxane

Samiha Sharlin, Rodrigo Lozano, Tyler R. Josephson

1,4-dioxane is an emerging water pollutant with high production volumes and a probable human carcinogen. The inadequacy of conventional treatment processes demonstrates a need for an effective remediation strategy. Crystalline nanoporous materials are cost-effective adsorbents due to their high capacity and selective separation in mixtures. This study explores the potential of all-silica zeolites for separation of 1,4-dioxane from water. These zeolites are highly hydrophobic and can preferentially adsorb nonpolar molecules from mixtures. We investigated six zeolite frameworks (BEA, EUO, FER, IFR, MFI, MOR) using Monte Carlo simulations in the Gibbs ensemble. The simulations indicate high selectivity by FER and EUO, especially at low pressures, which we attribute to pore sizes and shapes with more affinity to 1,4-dioxane. We also demonstrate a Monte Carlo simulation workflow using gauge cells to model the adsorption of an aqueous solution of 1,4-dioxane at 0.35 ppb concentration. We quantify 1,4-dioxane and water coadsorption, and observe selectivities ranging from 1.1×10^5 in MOR to 8.7×10^6 in FER. We also demonstrate that 1,4-dioxane is in the infinite dilution regime in both the aqueous and adsorbed phases at this concentration. This simulation technique can be extended to model other emerging water contaminants such as per- and polyfluoroalkyl substances (PFAS), chlorofluorocarbons, and others, which are also found in extremely low concentrations.

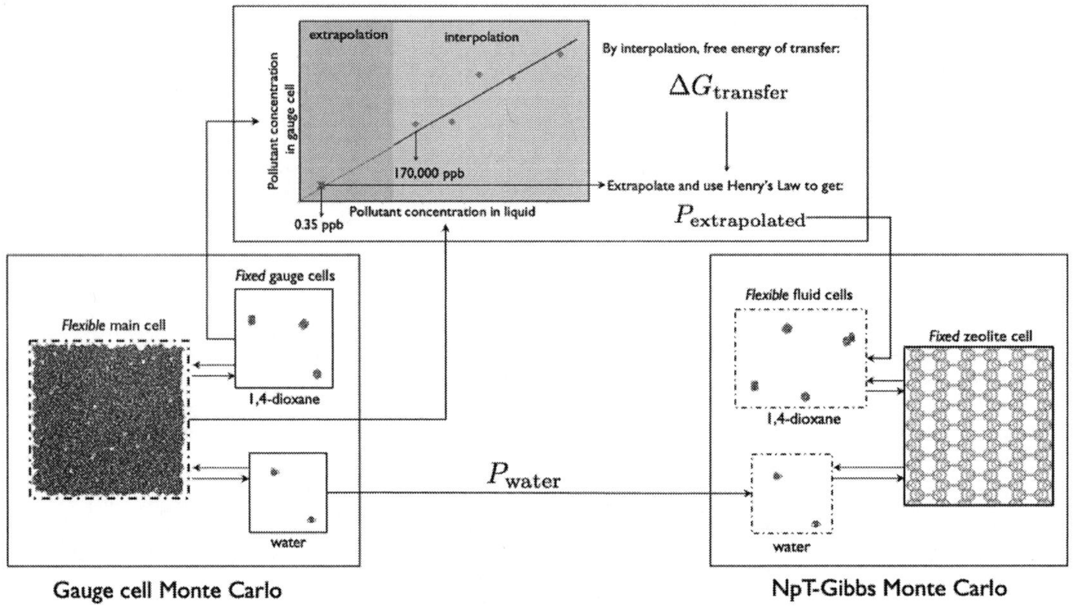

310h Deciphering the Molecular Modes Underlying NMR Relaxation Response in Fluids

Thiago Jose Pinheiro Dos Santos, Betul Orcan-Ekmekci, Dilip Asthagiri, Philip Singer, Walter Chapman

Nuclear magnetic resonance (NMR) relaxation is a powerful technique for probing matter non-destructively, with applications spanning medicine to material science to petrophysics. At high NMR frequencies, quantum effects are usually negligible, and a semi-classical statistical mechanical description is sufficient to model relaxation. Traditional theories of the NMR autocorrelation function for intramolecular dipole pairs assume single-exponential decay, while the actual autocorrelation of realistic systems display a rich, multi-exponential behavior resulting in anomalous NMR relaxation dispersion (i.e., frequency dependence). In this context, we have developed an approach based on "molecular modes of relaxation" to model and interpret the multi-exponential autocorrelation using simple, physical models within a rigorous statistical mechanical development. Our approach recasts the problem of evaluating the autocorrelation in terms of averaging over a diffusion propagator whose evolution is described by a Fokker-Planck equation. We have also shown that NMR relaxation of fluids can be probed within classical and semi-classical Molecular Dynamics (MD) simulations trajectories, which supports the validation of our theory and expands the range of practical applications. These molecular modes can be used to model and predict NMR dipole-dipole relaxation dispersion of fluids on the molecular level. It is worth noticing that both our statistical mechanical description of the distribution of molecular modes and our molecular dynamics simulations results assume no adjustable parameters.

310i Sequence-Encoded Spatial and Temporal Dependence of Viscoelasticity of Protein Condensates Using Computational Microrheology

Dinesh Sundaravadivelu Devarajan, Jeetain Mittal

Many biomolecular condensates exist as phase-separated viscoelastic protein assemblies within cells. Characterizing the viscoelasticity of condensates is essential for gaining insights into their spatiotemporal organization and physiological functions. Despite the significant interest in the dynamics and rheology of condensates, the quantification of their time-dependent viscoelastic properties is limited and mostly done through experimental rheological methods. Here, we demonstrate that a computational passive probe microrheology technique, coupled with continuum mechanics, can accurately quantify the linear viscoelasticity of condensates formed by intrinsically disordered proteins (IDPs) and is superior to other conventional computational rheology techniques. Using a transferable coarse-grained protein model, we first provide a physical basis for choosing optimal values that define the attributes of the probe particle, namely its size and interaction strength with the model IDPs. We show that the technique reveals the sequence-dependent elastic and viscous modulus, terminal viscosity, and relaxation time of heteropolymeric IDPs that differ either in sequence charge patterning in model IDPs or sequence hydrophobicity in naturally occurring IDPs. Further, we accurately quantify the spatial dependence of viscoelasticity in heterogeneous condensates formed by a pair of IDPs through spatially controlled probe motion in the microrheology technique. The computational passive rheology technique has important implications for investigating the time-dependent rheology of complex biomolecular architectures, resulting in the sequence-rheology-function relationship for condensates.

310j An Information-Theoretic Approach for Probing Macromolecular Phase Separation Via Data Compression

Ashley Guo, Gregory Dignon

A recurring challenge in molecular simulations is the selection of appropriate order parameters to describe complex molecular systems. An entropy-based descriptor derived directly from molecular coordinates would be especially valuable for studying assembly and aggregation processes. Recent work has introduced data compression as a novel approach to quantify order in both equilibrium and non-equilibrium systems and both identify and characterize phase transitions. While this has been applied toward both discrete and continuous particle models, as well as experimental data, the potential of this data compression approach for characterizing complex macromolecular systems remains unexplored. This work presents an initial application of the data compression methodology to identify and characterize phase transitions from molecular simulation trajectories of polymeric systems. Furthermore, we explore how this approach may be further applied toward the study of folding transitions in biomolecular systems and potentially manipulated to explore entropy as an order parameter when modeling complex molecular assemblies.

310k Navigating the Unknown: Efficiently Locating the Transition State of the Diels-Alder Reaction through Adaptively Sampled Point Clouds

Anastasia Georgiou, Bryan Munoz, Pranav Roy, Brandon C. Bukowski, Ioannis G. Kevrekidis

Locating transition states of chemical reactions is crucial for understanding reaction mechanisms and kinetics. Traditional methods such as metadynamics [1] and adaptive biasing force [2] require *a priori* knowledge of collective variables and exhaustive exploration to understand the potential energy surface; other methods, like the Nudged Elastic Band [3], require knowledge of good collective variables as well as of the minima to be connected. Here, we present an alternative approach, that does not require *a priori* knowledge of the collective variables, "Staying the Course," to locate the transition state of the Diels-Alder reaction in a data-driven fashion.

Our method builds upon previous work [5, 6, 7] in locating saddle points of stochastic dynamical systems without *a priori* knowledge of the manifold or collective variables. Instead, point clouds (positions of the atoms in the system) are adaptively & iteratively sampled along a 1D curve (here, an isocline) that drives the system from an initial conformation (typically a stable equilibrium) to a saddle point. When collective variables are unknown, we couple the algorithm with manifold learning techniques (diffusion maps) and Gaussian process regression to obtain the sought after path through local adaptive parameterizations of the effective free energy surface.

By integrating our algorithm with available atomistic and molecular simulation packages, we efficiently navigate the high-dimensional potential energy surface of an example problem (here, the Diels-Alder reaction). This approach allows for the automatic and adaptive identification of relevant reduced coordinates and the transition state without exhaustive sampling of the entire free energy landscape. Our results highlight the versatility of combining "Staying the Course" with molecular simulation codes for studying complex chemical reactions, showcasing its potential for applications in more complex reaction environments. We compare this to the metadynamics approach taken in [8].

[1] AD Bochevarov, E Harder, TF Hughes, JR Greenwood, DA Braden, DM Philipp, D Rinaldo, MD Halls, J Zhang, and RA Friesner, Jaguar: A High-Performance Quantum Chemistry Software Program with Strengths in Life and Materials Sciences, International Journal of Quantum Chemistry, 2013.

[2] W E, X Zhou. The Gentlest Ascent Dynamics. Nonlinearity, 2011.

[3] D Hoffman, R Nord, K Ruedenbergy. Gradient extremals. Theoretica chimica acts, 1986.

[4] J Yin, Z Huang, L Zhang. Constrained High-Index Saddle Dynamics for the Solution Landscape with Equality Constraints. Journal of Scientific Computing 2022, 91, 62.

[5] JM Bello-Rivas, A Georgiou, J Guckenheimer, and IG Kevrekidis. Staying the course: iteratively locating equilibria of dynamical systems on Riemannian manifolds defined by point-clouds. Journal of Mathematical Chemistry 61, 600–629 (2023)

[6] JM Bello-Rivas, A Georgiou, H Vandecasteele, and IG Kevrekidis. Gentlest ascent dynamics on manifolds defined by adaptively sampled point-clouds. The Journal of Physical Chemistry B 127, 5178-5189 (2023).

[7] A Georgiou, H Vandecasteele, and IG Kevrekidis. Locating saddle points using gradient extremals on manifolds adaptively revealed as point clouds. Chaos 33, 123108 (2023).

[8] CD Fu, LFL Oliviera, and J Pfaendtner. Assessing generic collective variables for determining reaction rates in metadynamics simulations. *Chem. Theory Comput.* 2017, 13, 3, 968–973

310l Transfer Learning of Symmetry-Adapted Learned Collective Variables from Classical to Ab Initio Simulations: Enhanced Sampling of Spin Defect Dynamics in Silicon Carbide

Soumajit Dutta, Cunzhi Zhang, Gustavo Perez Lemus, Juan J. de Pablo, Giulia Galli, Andrew Ferguson

Solid-state spin defects in silicon carbide (SiC) crystal lattices present promising qubit candidates with applications in quantum information science, sensing, and metrology [1]. The mechanisms and conditions for the generation, transport, and annihilation of these defects remains poorly understood. Improved understanding of these mechanisms can inform the realization of long-lived, stable qubits in an industrially relevant substrate. Accurate treatment of defect dynamics requires *ab initio* calculations accounting for the electronic structure, charge, and spin effects, but the high cost of these calculations typically precludes the direct simulation of these activated rare events, and direct application of enhanced sampling techniques to accelerate these processes is frustrated by a lack of knowledge of good collective variables (CVs) along with to bias sampling.

In this work, we report a strategy to learn CVs in inexpensive classical mechanical calculations and employ transfer learning to deploy them within *ab initio* enhanced sampling calculations. The underlying hypothesis is that the classical CVs can be easily learned and can provide a sufficiently good approximation for the many-body collective motions underpinning the defect dynamics to accelerate these processes within the *ab initio* simulations. Importantly, the indistinguishable nature of the C and Si atoms surrounding the defect means that the CVs capturing the defect transitions must respect the underlying translational, rotational, and permutational invariance of the underlying crystal. To this end, we employ our recently developed Permutationally Invariant Networks for Enhanced Sampling (PINES) approach for CV discovery that combines the Permutation Invariant Vector (PIV) featurization with regularized autoencoders to discover CVs that are, by construction, invariant to these symmetries [2].

We demonstrate our approach in applications to divacancy formation, reorientation, and antisite formation in SiC. We conduct classical CV discovery using PINES coupled to LAMMPS and then deploy the *ab initio* enhanced sampling calculations using the SSAGES enhanced sampling libraries coupled to the Qbox and CP2K simulation packages. As an additional outcome, we use the data harvested from our enhanced sampling calculations to parameterize new equivariant deep learning potentials for these processes in SiC.

References

1. E.M.Y. Lee, A. Yu, J.J. de Pablo, and G. Galli, Nat. Commun. 12 6325 (2021)
2. N.S.M. Herringer, S. Dasetty, D. Gandhi, J. Lee, and A.L. Ferguson, J. Chem. Theory Comput. 20 1 178-198 (2024)

115a Interfacial Selectivity of Peptide Surfactants at the Air-Water Interface for REE Recovery

Kathleen J. Stebe

We are developing peptide surfactants (PEPS) to capture and retain rare earth elements (REEs) at air-water interfaces to enable green recovery and separation of these critical elements. Selective separation of REE is particularly challenging among the lanthanide cations, which are present predominantly in a +3-oxidation state and have ionic radii that differ only by approximately 0.2 Å between the lightest and heaviest lanthanide. This contraction is exploited to generate selectivity among extractants used in liquid-liquid extraction (LLE). In this process, extractants in an organic phase (e.g. kerosene) are exposed to aqueous mixtures of REE cations. The extractants and REE cations form complexes at the aqueous-organic phase interface and subsequently partition into the organic phase. This process is weakly selective, with separation factors close to 1 for neighboring REE, so stages of LLE are required.

PEPS designed to bind selectively to particular REE cations are promising platforms for a new, green approach based on froth flotation. PEPS should bind selectively to form PEPS:REE complexes in bulk, and adsorb to interfaces of air bubbles sparged through solution for collection. Once adsorbed, each PEPS must retain its complexed REE, and cannot attract additional REE via non-selective Coulombic interactions that would compromise their selectivity. Here, we adopt lanthanide binding tags (LBTs) designed to bind selectively to Tb^{3+} as our PEPS platform. We use rheological characterization of LBT variants complexed with Tb^{3+} at air-water interfaces, supported by molecular dynamics and metadynamics simulations, to develop requirements for interfacial selectivity.

1. Interfacial Rheology of Lanthanide Binding Peptide Surfactants at the Air-Water Interface Stephen A. Crane, Felipe Jimenez-Angeles, Yiming Wang, Luis Ortuno Macias, Jason G. Marmorstein, Jiayi Deng, Mehdi Molaei, E. James Petersson, Ravi Radhakrishnan, Cesar de la Fuente-Nunez, Monica Olvera de la Cruz, Raymond S. Tu, Charles Maldarelli, Ivan J. Dmochowski, Kathleen J. Stebe, in preparation

115b Chirality and Complexity of Nanostructures

Nicholas Kotov

Chiral nanostructures are a large and rapidly evolving class of biomimetic materials. Besides fascinating optical, catalytic, and biological properties, the studies of chiral nanostructures represent a missing link in the emergence of complexity in Nature. *Complexity*, i.e. purposeful combination of order and disorder, leads to nontrivial combination of properties needed for many technologies. In the context of self-assembled particles, complexity manifests as spontaneous increase of structural hierarchy and correlated disorder seemingly prohibited by thermodynamics. These observations mirror the structural evolution of biological materials that combine nano-, meso- and microscale chirality. Taking an example of complex particles with twisted spikes and layer-by-layer-assembled nacre-like composites, we found that:

(a) formation of complex structures does not require monodispersity;

(b) competing thermodynamic restrictions in self-limited systems increase their complexity;

(c) synthetic particles can have higher complexity than their biological prototypes.

These findings were possible by applying graph theoretical (GT) measures of complexity to nanoparticle assemblies. Their applicability to 'imperfect' (nano)particles and ability to capture essential structural motifs made possible to include other nanoscale structures, such as complex porous particles and superlattices,[2] creating analogs of chemical formulas for complex particle systems (chiral, racemic and achiral). Understanding of intermolecular forces and expansion of self-limited assembly to bulk dispersions of diverse nano(particles) enabled the design of hierarchically organized bowtie-shaped particles with variable twist, size, and thickness and length.[3]

Structural complexity and technological significance are related. The simple pathways to complex particle systems with technological significance will be demonstrated for self-assembled chiral catalysts.[4] The complex particle nanosystems combining order and disorder with technological significance will be demonstrated for self-assembled analogs of cartilage that can be designed using GT for biomedical devices and batteries [5,6].

References

[1] W. Jiang, et al., Emergence of Complexity in Hierarchically Organized Chiral Particles, *Science*, **2020**, 368, 6491, 642-648.

[2] S. Zhou, et al, Chiral assemblies of pinwheel superlattices on substrates, *Nature,* **2022,** 612, 259.

[3] P. Kumar, et al. Photonically Active Bowtie Nanoassemblies with Chirality Continuum, *Nature,* **2023**, 615, 418–424.

[4]. S. Li et al Single- and Multi-Component Chiral Supraparticles As Modular Enantioselective Catalysts, *Nature Comm,***2019**, 10, 4826.

[5] Wang M. et al., Biomimetic Solid-State Zn^{2+} Electrolyte for Corrugated Structural Batteries, *ACS Nano*, **2019**; 13(2), 1107. PMID: 3060811

[6] Wang M. *et al.*, Biomorphic Structural Batteries for Robotics. *Sci. Robot.* **2020**, 5; https://doi.org/10.1126/scirobotics.aba1912.

115c Making the Most of Molecular Methods

Greg van Anders

Molecular simulation methods have solidified as a foundational tool in chemical engineering and adjacent areas of science and engineering. However, chemical engineers have also shown, particularly through the fundamental contributions of Sharon Glotzer, that molecular methods generalize and provide equally strong utility for systems in which the 'molecules' actually represent nanoparticles or colloids. How far can this program can be taken? And what kinds of insight could molecular methods provide for other, even 'less molecular' systems? Here, I will give examples of applications of molecular methods to problems in additive manufacturing and neural network training where the systems in question bear little initial resemblance to molecules. Moreover, I will show that, despite the apparent differences in system construction, molecular methods deliver powerful forms of insight that go beyond conventional approaches.

115d Sequence-Dependent Phase Separation and Aggregation in Model Proteins

Athanassios Panagiotopoulos

Biomolecular phase separation has emerged as an organizing principle underlying subcellular organization and controlling many processes in living systems. The time and length scales relevant for phase transitions of biomolecules far exceed those accessible thought atomistic simulations, so coarse-grained, physics-oriented models are appropriate. A specific principle of molecular organization turns out to be competition between macroscopic phase separation and formation of finite-size aggregates, which is also relevant for micelle formation by surfactants. Recent experiments on synthetic polymers with controlled sequences of solvophobic and solvophilic blocks also display this competition between aggregation and phase separation. In addition, very large, nearly monodisperse equilibrium aggregates are observed at specific thermodynamic conditions in these systems. Grand canonical Monte Carlo simulations shed light on the sequence dependence of the phase separation behavior and point to a novel 'surface enrichment' microscopic mechanism for formation of equilibrium aggregates that likely has biological relevance.

115e Threading the Thermodynamic Needle - Balancing Enthalpy and Entropy to Design Complex and Reconfigurable Assemblies

Thi Vo

Designing nanoscale synthons with predictive control over shape, size, and interparticle interactions is a holy grail for materials design. To date, there exists a large space of experimental parameters that can be employed to direct nanoscale assemblies. Tuning such experimental handles produces different modes of interactions such as hard particle packing, soft deformable ligand shells, or patchy/selective attraction. However, a major limitation intrinsic to such building blocks lies in the fixed nature of their pre-programmed interactions. This means that the assembled morphologies are also often static, thereby limiting the *a priori* design of reconfigurable structures. Here, we present strategies for designing building blocks that address the above limitation. We first discuss theoretical insights into how to imbue secondary interactions into synthesized building blocks that can be toggled on/off via experimentally relevant external stimuli such as temperature, external field, and/or solvent change. We then showcase how synthon reconfigurability shifts the preferred, thermodynamically stable assembly structures, validated by both simulation and experiments. Our works provide unique insights into nanoscale synthesis, allowing for *a priori* design of complex building blocks that can target novel and reconfigurable assemblies.

115f Shape Dependent Ordering of Hereditary Spherocytosis Inspired RBCs

Omolola Eniola-Adefeso

Introduction:

Hereditary spherocytosis is a genetic disorder that causes red blood cells (RBCs) to become more rigid and spherical in shape. This apparent change in RBC morphology leads to lifelong health complications such as anemia, jaundice, gallstones, spleen damage, and potential loss of life without surgical intervention. Current clinical diagnostic methods can be prohibitively expensive and nevertheless, due to the mildness of most cases the disorder is not diagnosed. By leveraging the well-studied hexactic packing of spheres, it may be possible to identify structures caused by spherocytes using a minimal cost sedimentation method. This study seeks to quantify the granular structures for mixtures of discocytes (biconcave disc RBCs) and chemically altered spherocytes (spherical RBCs) in a sedimentation system.

Materials and Methods:

Blood draw protocols have been approved by the University of Michigan Internal Review Board (IRB-MED). RBCs were separated from whole blood samples using centrifugation with the plasma and white blood cells being discarded. Spherocytes were made from healthy RBC samples via exposure to sodium salicylate. Rigidification of RBCs was produced through exposure to tert-butyl hydroperoxide. Rigid spherocytes were made through exposure to both of the above. Alteration of RBC shape was confirmed with confocal imaging of fluorescently stained RBC membranes via Wheat Germ Agglutinin, Alexa Fluor™ 488 (Thermo Fisher). Fractions or pure samples of discocytes and spherocytes were then reconstituted in phosphate buffer solution at less than 1% hematocrit. The blood solutions were then pipetted onto the bottom side of a suspended glass slide and allowed to sediment for 10 minutes. The monolayer of RBCs formed at the air-water boundary where then imaged using light microscopy. Centroid locations of each cell were selected via ImageJ. Centroids were analyzed using OVITO software to quantify average hexactic ordering, radial distribution functions, and clustering of cells.

Results and Discussion:

We describe the radial distribution functions of each cell type, hexactic order, clustering of cell mixtures using advanced analysis computational methods. We find that the presence of spherocytes can cause differences in the ordering and disordering of the sedimentation system. We investigate the role of rigidity and sphering in the differential hexactic ordering in unary systems, and find shape plays a more significant role. We find that spherocytes produce higher long-range order than discocytes. We also investigated binary fractions of discocytes and spherocytes. We find the onset of aggregation of each cell type occur above a fraction of 0.5. This aligns with a random placement simulation on a hexatic lattice, which indicates that sedimentation is likely random. Increases in hexatic order as a function of cell type fraction are more significantly driven by the aggregation of spherocytes. Likewise, global long-range ordering is also driven significantly after the onset of spherocyte aggregation.

Conclusions:

This work details how cell shape and rigidity can impact the ordering of red blood cells (RBCs) that have sedimented at an interface without additional external forces. We complete the first ex vivo quantification of RBCs hexatic ordering. Using artificially sphered and rigidified RBCs, inspired by hereditary spherocytosis, we isolate the ordering effects of shape and rigidity. We find that shape plays a significant role in the enhanced local and long-range ordering of RBCs in unary and binary systems. This work highlights how an interdisciplinary lens of routine soft matter analysis and common biological phenomenon can symbiotically produce meaningful results. This work advances granular dynamics of RBCs and lays a foundation for future studies into bio-granular phenomenon for other shape changing RBC diseases. Future works include the inclusion of patient samples, automation, and bridging towards other types of cells.

115g Integrating Experiment and Simulation to Understand and Control the Structural Color of Defect-Laden Colloidal Crystals

Michael Solomon

Colloids self-assembled into crystals can recapitulate the iridescent structural color found in organisms such as butterflies, beetles, and cephalopods. However, the brilliance of the color in artificial materials often falls short of what is achieved in living systems. One origin of this discrepancy are defect structures present in the artificially self-assembled materials; such defects are typically abundant in colloidal crystals. Here we discuss work to understand and anneal defect structures in colloidal crystals; we link the type and abundance of defects directly with the intensity of the material's structural color response. We describe how this progress was accomplished by close integration of confocal microscopy experiments and computer simulation. Furthermore, and in honor of Sharon C. Glotzer, we discuss how the experimental component of the self-assembly and optical characterization has benefited from this integration, because using simulation methods not only provides mechanistic understanding of confocal microscopy observations, but also expands the range of conditions that can be analyzed.

115h The Shape of Molecular Interactions: Using Classical Models to Rationalize Atomistic Behavior

Rose Cersonsky

Entropy is a driving force of self-organization, as witnessed by the many instances of self-assembly occurring in systems without explicit enthalpic interactions. However, how can we understand this driving force when enthalpy is also at play? In honor of Sharon Glotzer's 60th birthday, I will revisit some earlier work to discuss the role of entropy in atomistic systems, from high pressure group IV elements to metals to organic molecules, weaving a line as to how entropy can be understood in complex systems, not only those where we set U=0.

115i Data-Driven Backmapping of Coarse-Grained Simulations of Proteins and DNA

Andrew Ferguson

Data-driven modeling and deep learning present powerful tools that are opening up new paradigms and opportunities in the understanding, discovery, and design of soft and biological materials. I will describe approaches based on denoising diffusion models and flow matching to restore all-atom resolution to coarse-grained molecular dynamics simulations of proteins and DNA to help break the time scale barrier in biomolecular simulations.

115j Divergence between Structure and Nucleation Outcome in Heterogeneous Nucleation of Close-Packed Crystals

Amir Haji-Akbari

Crystallization is a process in which molecules and/or building blocks within a melt or a solution assemble into a crystalline phase, and can occur in a wide variety of atomic, molecular, metallic, polymeric and colloidal systems over a wide range of time and length scales. Crystallization almost always proceeds through a process known as nucleation and growth wherein a sufficiently large crystalline seed emerges during nucleation, which then grows by absorbing more building blocks until the system reaches thermodynamic equilibrium. Under most circumstances, nucleation occurs heterogeneously in which an external surface facilitates crystallization by decreasing nucleation barriers. Despite being the dominant mechanism of crystallization and self-assembly, there are major gaps in our understanding of how surface chemistry and topography impacts the kinetics and mechanism of heterogeneous nucleation.

In recent decades, molecular simulations have emerged as attractive tools for filling this gap and have been successfully used for probing the physics of nucleation in wide variety of systems. One of the necessary ingredients of such explorations is the construction of good collective variables (CVs) that accurately describe the progress of nucleation. Historically, such order parameters have been developed by comparing the static structures of bulk target crystals and the corresponding fluids [1].

In this talk, I will discuss our recent work [2] in which we use molecular dynamics (MD) simulations, jumpy forward flux sampling (jFFS) [3] and machine learning to probe heterogeneous nucleation of close-packed crystals in the Lennard-Jones and hard sphere systems. We demonstrate that traditional CVs based on structural properties of bulk crystals fail at describing the progress of heterogeneous nucleation despite being superb descriptors of homogeneous nucleation of the same crystals. There is therefore a divergence between traditional notions of crystalline structure and nucleation outcome, which is accompanied by an intriguing dynamical anomaly wherein crystalline particles with few crystalline neighbors tend to move faster than their liquid-like counterparts. Through high throughput screening of hundreds of thousands of CVs and machine learning, we systematically assess different strategies to refine such CVs. In particular, we find readjusting Steinhardt bond order parameter cutoffs and pruning crystalline bridges to be effective strategies in improving CV efficacy. Employing such optimal CVs is pivotal for proper characterization of heterogeneous nucleation in metallic and colloidal systems that typically form such close-packed crystals.

[1] Steinhardt PJ, Nelson DR, Ronchetti M, Phys. Rev. B, 28: 784 (1983).

[2] Domingues TS, Hussain S, Haji-Akbari A, J. Phys. Chem. Lett., 15: 1279 (2024).

[3] Haji-Akbari A, J. Chem. Phys., 149: 072303 (2018).

AUTHOR INDEX

AIChE
120 Wall Street, FL 23
New York, NY 10005-4020

ISBN 979-8-3313-1656-3

CURRAN ASSOCIATES INC.
proceedings.com